TRANSLATING STATISTICS TO MAKE DECISIONS

A GUIDE FOR THE NON-STATISTICIAN

Victoria Cox

Translating Statistics to Make Decisions: A Guide for the Non-Statistician

Victoria Cox
Salisbury, United Kingdom

ISBN-13 (pbk): 978-1-4842-2255-3   ISBN-13 (electronic): 978-1-4842-2256-0
DOI 10.1007/978-1-4842-2256-0

Library of Congress Control Number: 2017934647

Managing Director: Welmoed Spahr
Editorial Director: Todd Green
Acquisitions Editor: Robert Hutchinson
Development Editor: Laura Berendson
Technical Reviewer: Alison Berry
Coordinating Editor: Rita Fernando
Copy Editor: Judy Ann Levine
Compositor: SPi Global
Indexer: SPi Global
Artist: SPi Global
Cover image designed by eStudio Calamar

Distributed to the book trade worldwide by Springer Science+Business Media New York, 233 Spring Street, 6th Floor, New York, NY 10013. Phone 1-800-SPRINGER, fax (201) 348-4505, e-mail orders-ny@springer-sbm.com, or visit www.springeronline.com. Apress Media, LLC is a California LLC and the sole member (owner) is Springer Science + Business Media Finance Inc (SSBM Finance Inc). SSBM Finance Inc is a **Delaware** corporation.

For information on translations, please e-mail rights@apress.com, or visit http://www.apress.com/rights-permissions.

Apress titles may be purchased in bulk for academic, corporate, or promotional use. eBook versions and licenses are also available for most titles. For more information, reference our Print and eBook Bulk Sales web page at http://www.apress.com/bulk-sales.

Any source code or other supplementary material referenced by the author in this book is available to readers on GitHub via the book's product page, located at www.apress.com/9781484222553. For more detailed information, please visit http://www.apress.com/source-code.

Apress Business: The Unbiased Source of Business Information

Apress business books provide essential information and practical advice, each written for practitioners by recognized experts. Busy managers and professionals in all areas of the business world—and at all levels of technical sophistication—look to our books for the actionable ideas and tools they need to solve problems, update and enhance their professional skills, make their work lives easier, and capitalize on opportunity.

Whatever the topic on the business spectrum—entrepreneurship, finance, sales, marketing, management, regulation, information technology, among others—Apress has been praised for providing the objective information and unbiased advice you need to excel in your daily work life. Our authors have no axes to grind; they understand they have one job only—to deliver up-to-date, accurate information simply, concisely, and with deep insight that addresses the real needs of our readers.

It is increasingly hard to find information—whether in the news media, on the Internet, and now all too often in books—that is even-handed and has your best interests at heart. We therefore hope that you enjoy this book, which has been carefully crafted to meet our standards of quality and unbiased coverage.

We are always interested in your feedback or ideas for new titles. Perhaps you'd even like to write a book yourself. Whatever the case, reach out to us at editorial@apress.com and an editor will respond swiftly. Incidentally, at the back of this book, you will find a list of useful related titles. Please visit us at www.apress.com to sign up for newsletters and discounts on future purchases.

The Apress Business Team

*To my parents, Jacquie and Andrew,
who have supported me through everything and
without whom this would not have been possible.*

Thank you.

Contents

About the Author

Victoria Cox is a senior statistician at the United Kingdom's Defence Science and Technology Laboratory (Dstl). There she advises internal teams and external organizations, both nationally and internationally, on the experimental design of statistical studies and the application of the methods of statistical analysis to a wide variety of practical problems. She teaches statistical courses to non-statistician scientists, technologists, analysts, managers, and executives. Cox took her degree in mathematics from the University of Sheffield and studied at l'Ecole Nationale de la Statistique et de l'Analyse de l'Information (ENSAI).

About the Technical Reviewer

Alison Berry has an honors degree in Statistics from the University of Reading and is accredited as a Chartered Statistician by the Royal Statistical Society. She has worked in the United Kingdom's Ministry of Defence for 20 years as a senior statistician and an analyst.

Introduction

This introduction will give an overview as to what the book is about, who the author—me, is, how the book is laid out, what you should learn by its end, and more information about R - if you choose to use it.

What the Book Is About

This book is designed to highlight typical misconceptions non-statistics users can have, to show common pitfalls that (new) statisticians and those dipping their toes in the water can fall into, and to suggest solutions to correct and avoid them.

Although the book is written from my viewpoint and therefore the statistician's viewpoint, the emphasis is on translating the output correctly to aid evidence-based business decision making.

Statistics can seem like a tricky subject at first, but once you can get your mind around the basic ideas, the rest will come easily. The general process, which we will walk through in the book, has a clear order from designing an experiment, through analysis, to reporting the results. However once understanding has been achieved in certain terms and ideas, such as confidence intervals, p-values, and so forth, this can make the more complicated methods simpler. For example, some of the more complex models are just extensions of simple hypothesis tests in terms of the output and the results you would pull out.

Once immersed in the world of statistics it is very easy to get lost in the language - however this book is written with the non-statistician customer in mind. Don't worry, there won't be any complicated algebra or proofs included as there are plenty of other texts for that purpose.

The key to effective decision making is to make full use of all available relevant data. However if this is not in an accessible language then sometimes important information can be ignored. This is where being able to translate the statistical jargon into English is vital, and this can be a whole new skill in itself. Throughout the book we discuss how to make reporting the results an easy and understandable task.

Each section contains advice about how to proceed with each type of data, method, and so forth along with potential pitfalls and misconceptions related to the topic. There is dummy example data provided, however the case studies they refer to in the majority of cases are real scenarios. R code is provided to help you run through the examples should you wish, but it is not necessary that you have previous knowledge of R. The text is written in such a way that the example code is extra information to that which is already provided.

The main aim throughout is to make sure the right methods are being used on the right data, so the right output is produced and the right conclusions are translated to the customer. We want to avoid ending up giving wrong or misleading conclusions to the customer due to something going awry at one of the many points throughout the design and analysis process.

Who Is the Author?

I am a senior statistician, so day to day I am absorbed in consulting on statistics, creating experimental designs, and conducting analysis. However I work for a defense organization where most of my customers and colleagues have never done statistics, or have done very little, so it is my job to help them understand the messages in their data. A lot of statistical output can be produced, but it would be unproductive for them to see it all. The challenging task is to pick out the important information, which should include main assumptions and "negative" results, and then condense it to a manageable size.

My day job involves working with military, industry, academic, and government colleagues and customers. Part of a statistician's task is to learn about the subject matter so the results can be translated accordingly, which is why I made the choice of having a wide variety of case studies in the book.

I have never been a fancy wordsmith, but that has actually been extremely beneficial in being able to simplify statistical terms to those who haven't studied the subject. This also has helped with the internal Introduction to Statistics course I run for those who want to improve their statistical ability, as I've found I can teach the course without people being completely baffled by the, quite frankly, backward statistics language.

How Is the Book Laid Out?

Presumably you will have read the contents page, but I will go into a bit more detail about each of the sections. At this point there may be terms you are unfamiliar with, don't worry, these will be explained in the relevant chapters.

There will be examples and figures included throughout the book; many of the examples will be run in R, but for those not wishing to use R there are sections translating the output at each stage with summaries at the ends.

Chapter 1: Design of Experiments: What Do I Need to Do to Get the Data?

Every experiment, study, or trial needs to begin with design of experiments (DoE), which may sound logical; however it isn't always followed in practice. Even when there is some type of design planned, people sometimes cut corners, either intentionally or unintentionally. Even if there are constraints on the number of repeats that can be run or available time, the experimental design phase should still be undertaken and it can account for these limitations. This chapter investigates: establishing the point of the study; thinking about the variables, including interactions and confounding; achieving the required power and confidence levels; and determining the appropriate designs, for both physical and computer experiments. It also briefly looks at survey design.

Chapter 2: Data Collection: How Do I Get the Data?

Many issues can arise with data collection and some are unforeseeable. However there are ways of minimizing these as much as possible. There also can be the added complication of data formatting, and depending on the planning that was put in place initially, can determine the amount of time spent on this theme. This chapter looks into both of these areas and some of the methods that can help avoid such things as missing data and time being wasted.

Chapter 3: Exploratory Data Analysis: What Data Do I Have?

Once the data has been collected you shouldn't go straight into analysis, there is a step before to make sure you're on the right analysis path and that's exploratory data analysis (EDA). EDA is about getting to know the data you have, which mainly involves visualization techniques, so you know what you can and can't proceed to do with the data. This chapter explores the pieces that make up EDA: identifying different data types, viewing the data with basic plots, detecting potential outliers, discovering the data distribution, and revealing possible trends.

Chapter 4: Descriptive Statistics: What Can the Data Tell Me?

Descriptive statistics is a good place to start with the initial numerical investigation of your data, as your data is only a sample from a larger population. This stage is still in relation to learning more about your data before conducting any types of hypothesis tests. There are many descriptive statistics that will look at the location, dispersion, and shape of continuous data along with a smaller number of methods for investigating discrete data, and we delve into them in this chapter.

Chapter 5: Measuring Uncertainty: How Good Is the Data?

No analysis will ever give you an exact answer. Unless you can test the entire population, there will always be some inherent uncertainty in the results. There are multiple ways to determine uncertainty depending on the focus of the overarching question, and this chapter looks at some of the most common including confidence intervals, tolerance intervals, and prediction intervals along with some examples.

Chapter 6: Hypothesis Testing: What Differences Are in the Data?

Though the official statistical language and meaning is all backward, for simplicity we say that hypothesis testing is used to find whether differences are likely due to chance or likely due to a significant difference. This chapter discusses "simple" hypothesis testing, and by simple it just means looking at one explanatory variable using tests, such as t-tests, proportion tests, and nonparametric equivalents. It is split into two main sections, a section that details the components that make up hypothesis testing, such as p-values, differences (significant, practical, etc.), and a section that looks at some of the available tests given specific data types.

Chapter 7: Statistical Modeling: What Is Actually Going On in the Data?

Once there starts to be more explanatory variables in your study, the analysis needs to move up to statistical modeling, though generally the same theories from simple hypothesis testing hold. Models can be used to test the importance of different variables, predict future outcomes, or to assign uncertainty to the results. This is the largest chapter of the book as statistical modeling is a large part of statistical analysis. The chapter is split into two main sections, the model components (i.e., the process to follow, common outputs, etc.) and some of the more common types of statistical models themselves. There are multiple examples in the second section as one of the easiest ways to understand and translate model outputs is to go through examples.

Chapter 8: Multivariate Analysis: What Have I Found in My Larger Data?

When you begin to have more response variables you need to dig into the data using multivariate methods. This data is generally "large" however it isn't necessarily as large as that which the term *big data* represents. There are numerous multivariate techniques a lot of which overlap and as such not all could be shown. Within this chapter we inspect three of the most common methods appropriate for this type of data (whilst being the most different to each other): MANOVA, PCA, and Q methodology. There is guidance in how to start reading the output as it's not such a straight forward task, as with statistical modeling output.

Chapter 9: Graphs: What Does the Data Look Like?

Graphing results is an easy way to emphasize the key messages that the analysis output indicated; as long as the plots are kept clear and concise. Although we touched on viewing the data in the EDA chapter, this chapter probes deeper into general advice about the pitfalls to avoid, highlights how to change the aesthetics in plots created in R, and then shows how to create some of the common plot types.

Chapter 10: Translation and Communication: How Do I Get the Message Across?

Being able to translate the statistical jargon into understandable language is, as I mentioned previously, a skill all by itself. Throughout the book I have tried to demonstrate ways to convert the output into an understandable message that can be used for effective decision making. This final chapter aims to tie this all together along with guidance on what information to include, where it's best placed, how to structure the message, and what other considerations need to be considered. It also provides four examples drawn from previous chapters to show what information should be picked from the output to present to the customer, and how to do this clearly and efficiently.

What Will Be Learned?

In an ideal world by the end of this book you shouldn't need it anymore. However in reality there is a lot to remember and this will be a good guide to keep handy for reference of good practice in statistics. We only have a finite brain storage capacity, and if statistics is not something you use daily then it's easy to forget and start slipping into those pitfalls again. It doesn't even matter how much of a veteran in statistics you are, it can be easy to make mistakes, especially in new fields and this can serve as a refresher. Although I don't claim to cover everything, it would be a much larger book for a start, this is a good basic book of the most common topics non-statisticians and beginners alike would use. Even then some of the advice will be relevant for more specific complex studies, for example, DoE, EDA, graphs, translation.

The aim of this book is to be a guide for recognizing potential mistakes and learning techniques for avoiding them or overcoming them at each section of the study. It should lead you down the correct path of thinking and make sure that you have thought about each stage progressively so in the long run time and money won't have been wasted and the results presented will be meaningful.

The whole reason we carry out statistics is to inform decision making so we need to make sure we produce useful, reliable, defendable results, and working in defense I know how serious getting those key messages wrong could be. Along with this we also need to make sure the associated assumptions and uncertainty is understood.

If you chose to try out R, download for free from https://cran.r-project.org/, then this book is also an initial introduction to some of the code and packages available. The data will be included at the start of all the examples and the R code will include instructions detailing the use of each line. By the end of the book you will be able to refer back to the appropriate code to use on your own data and understand which bits you need to tweak to make it work for your data.

If you are interested in using R then the next section will give more details, otherwise you can skip ahead to Chapter 1.

Using R

At the time of writing this book, the R version used was R v3.3.1, and as such the packages listed and code may need to be tweaked for future versions.

If you have never used R before it can seem quite daunting, as when you open it up, there is only a mostly blank box. Don't worry, that's just where you will be copying and pasting the code I have created; it's also where the output will be displayed.

Whenever there is a new package listed, for example **MASS**, you will need to click on the menu bar: Packages/Install package(s)..., you then need to choose a "mirror"—one that is local to you. Once that has been done a long list of packages will be shown, and you need to find the one needed, here **MASS**. Once that has been done you just need to run the R code as usual, and when you get to *library(MASS)*, it will load the library into your session. Note that once you have installed a package from the list, you only ever need to run the *library()* command to load it, you won't need to install it each time unless you are using a different computer or a new version of R. All packages and R commands that are in the body text of the book will be shown in the same font as used above.

There also will be dependencies for certain packages, but these should automatically be installed once you have installed the main package. If there is any command listed that you would like more information on you can type *help()* and put the command within the parentheses, for example, *help(pwr.t.test)* for the *pwr.t.test()* command in Chapter 1.

If you want to preemptively install the R packages, then the list of packages to be used throughout the book are: *pwr, AlgDesign, Rmisc, Hmisc, tolerance, RcmdrMisc, exact2x2, ggplot2, car, PairedData, MASS, lsmeans, brglm, pscl, AER, HH, lmerTest, mvnormtest, biotools, mvoutlier, RVAideMemoire, shotGroups, corrplot, FactoMineR, factoextra,* and *qmethod*.

As a side note, R does not like the quotation marks that Word creates, so while every effort has been made to replace these with R friendly versions the odd one may have been missed. Therefore, if you get an error this may be one of the reasons why.

In some cases the plots in the examples may not match the R code provided, they may show just the basic plot, and this is purely for space saving in the book. Chapter 9 explains how to add details in **ggplot()** so if you wanted to go back to try to replicate the graphs in previous chapters you can. In addition the plots were created to include color, which will be shown in the electronic version, however they won't be in the print version; although the output in R will show you the expected colors.

The examples using R comprise four components: a line explaining what the R code is about to do, which always starts with the number symbol (#); the R code itself; a print out of the results; and an explanation of what the results actually mean. Any lines beginning with # can be entered into the R script window as it recognizes it as text that isn't an R command.

Design of Experiments

What Do I Need to Do to Get the Data?

> To call in the statistician after the experiment is done may be no more than asking him to perform a post-mortem examination: he may be able to say what the experiment died of.
>
> —R. A. Fisher[1]

As Sir Ronald Fisher so eloquently put it, if an experiment isn't designed well then any conclusions made from the resulting data may not be useable or may have to be reported with too many accompanying caveats.

The point of an experimental design is to systematically investigate the problem space to a required minimized risk level, basically so you get the most out of your relevant collected data. However the main thing to note here is that it doesn't take long to design a good experiment, you just need to think about in the right order.

Figure 1-1 shows the suggested process to follow when thinking about designing any experiment with the following sections delving into more detail for each part of the process.

---

[1]Presidential Address to the First Indian Statistical Congress, 1938.

© Victoria Cox 2017
V. Cox, *Translating Statistics to Make Decisions*, DOI 10.1007/978-1-4842-2256-0_1

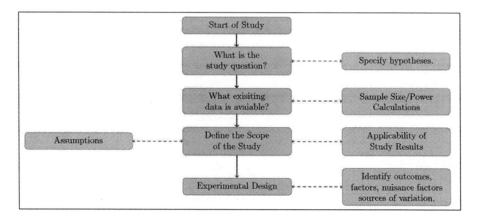

Figure I-I. Design of experiments thought process

Straight away you can see that experimental design is more than just a question of "how many do I need to do?" That is a common misconception. With that in mind, let's go through the steps in the process.

Forming the Study Question

The first item that needs to be done is to actually define the question. Generally the customer will give you a task such as "how good are these new detectors?" Then the initial stage of the process is to establish what they really want to know and why.

You need to gain as much (sensible) information as possible so that the experiment doesn't go down the wrong route and the resulting data will be able to adequately answer the customer's question. Knowing the key questions to ask can facilitate the discussion and help the customer understand why this information is important. It also ensures that you are both on the same page about what is required, and how the project will progress.

Forming Hypotheses

The reason for delving into the exact study question is due to the fact that it helps form the hypothesis to be tested in the analysis, an in depth description can be found in Chapter 6. Briefly, you have a null hypothesis, which is your "no action" case and your alternative hypothesis, which is your "take action" case. Therefore with a customer question of is the new method better than our own, the null hypothesis would be there is no difference between the current and new methods and the alternative would be that the new method is better than the current with the latter requiring things to be changed to use the new method, hence action case.

Hypotheses also can be either one-sided or two-sided. This just means that for a one-sided hypothesis you are interested in your sample being either larger or smaller than a value or another sample but not both. Two-sided means you are interested in a general difference between samples or between the sample and a threshold value regardless of direction.

This information about sidedness is important as it has an effect not only on the analysis but also on the power and sample size calculations as the risk calculated for each is split differently, more in Chapter 5. Generally, if the interest is only one-sided then a smaller sample size will be required as there will be no information about the other "side" of the data. However, it is important to note that this needs to be decided before the experiment and not changed afterward, otherwise the design of experiments may not be suitable.

Information Required

A lot of the time the design of experiments process needs to be iterative due to resource constraints, such as having limited funds or time, but here are some key points to discuss up front:

- *What is the question you want answered?* This can help lead you to form a hypothesis and also to define the other questions.

- *Why does it need answering?* By getting more detail it can actually change the emphasis of the question they want answered and should focus the design and testing that will be required (e.g., "are the new detectors better than our current detector and which ones are best?"). It also can help highlight any subquestions.

- *What do you want investigated?* Asking this question will enable the customer to think about the variables that will be used in the design along with the numerical ranges of interest. It also could uncover any variables that may have been overlooked, see the Experimental Design section.

- *How are the results going to be applied?* This can lead to thoughts about applicability to other areas as well as highlighting the assumptions that need to be made—more in the Defining the Scope of the Study section.

- *What level of risk is acceptable?* You want to know about the confidence and power levels. Translating this information into risk, however, can be much more accessible for non-statisticians, see more detail in the Power and Sample Size section.

- *Can you explain some of the subject matter to me?* This bullet may be optional dependent on your background - if the subject matter is new to you. By getting more details about the practical side allows you to comprehend what is realistically doable in terms of conducting the experiment (e.g., applying complete randomization to an experiment that requires 30 min. to set up each iteration is not practical).

You need to start with the broad questions then lead into the more precise questions, such as those about risk levels, to get the answers you need to continue. Once you have all the information required and have moved to the process of using any existing data, it can soon highlight issues about achieving the required risk levels due to possible constraints or large variation. This would necessitate further discussions about the scope of the study and the variables themselves—hence the process being an iterative one.

Power and Sample Size

Before looking at the information needed as well as how to actually run the calculations there are some misconceptions that need to be addressed.

Contrary to popular belief bigger is not always better, at some point the relative gain will plateau off, therefore doing more repeats would be a waste of resources. The sample size will vary depending on the type of data you are looking at, the variation of the data, and what risk levels you need to achieve.

On the other hand, it is wrong to use a sample size of 3 just because you could get published. Many non-statistical journals are now improving their review process by demanding a power calculation be included in the paper, which is good practice. In terms of animal studies, although we all must adhere to using as small a number of animals as possible, there is no point running an experiment if the power is weak as that means the results will be worthless and therefore a waste of those lives. It would be better not to run the experiment rather than gain inconclusive results just from trying to look as if you are using fewer animals. The ethical review process requires power calculations to be included, as under-powered proposals will be rejected due to that last point.

Another common misconception is about significance; generating a larger sample size will not guarantee significant results. If there are significant effects to be found then an adequate sample size will be efficient for finding them. However, there always will be your defined level of risk that they aren't found, which is one minus the power level—see the Risk section. There is also the fact that there may not be any significant differences to find in the first place.

A significant difference may not be a practical difference and this can occur when the sample size is overly large. If the data you are collecting has very little variation and you have arbitrarily chosen to do 1,000 repeats, you may find a significant difference of no importance. Say the 1,000 repeats were to compare the top two marathon runners' completion times; you may find a significant difference of 5 seconds. Is that data really useful? Also is it practical that the runners have to complete 1,000 marathons? There will be a practical difference that is of interest and this information will be used in the calculations, but it also can be stated in the analysis if something like the this occurs: e.g., the analysis showed that A runs significantly quicker than B, however this is a statistically significant difference and not a practical difference due to the fact that it is only on average by 5 seconds.

Calculation Information

When conducting a power or sample size calculation there is certain information required that you will need to get from the customer. The type of information depends on the type of data you will be collecting and what you are looking to obtain. Here I look at simple examples of the two most common types: means from continuous data and proportions from binary data. First I explain risk in terms of confidence and power as that is required in both example cases, in fact at least one of the two risks will be required in all calculations.

Risk (Confidence and Power Levels)

For any calculation you will require a confidence level, a power level, or both, depending on whether you require a power calculation, a confidence level, or a sample size. If you have a set sample size due to resource constraints then you won't need to set one of the power or confidence levels, but you will still need to understand what they mean. Figure 1-2 shows a table of the possible errors that any experiment can have.

		Findings in Experiment	
		Evidence of a significant difference	No evidence of a significant difference
Truth in Population	No difference in population	Type 1 error = α Confidence = $1 - \alpha$	No error
	Difference in population	No error	Type 2 error = β Power = $1 - \beta$

Figure 1-2. Possible errors in an experiment

The labels across the top show the results found in the experiment and the labels down the left show the truth if the entire population could be tested. There are clearly the situations where the experiment matches the population truth (shown in green), which is what we are aiming for by reducing the chance of the other errors.

In the top left box of the four colored boxes, if we find a significant difference when there shouldn't have been one in reality, that's called a Type 1 error. This is where the terms *confidence* and *significance* are associated. They are one minus each other, e.g., 95% confidence is equivalent to 5% significance. If we think about our detector example it may be that we find a spurious difference between similar detectors and unnecessarily spend money on the new detector, so here we have found something we should not have and wasted money.

In the bottom right box of the four colored boxes, if we don't find a significant difference when there should have been one in reality, that is called a Type 2 error. This is where the term *power* is associated. There is no official term for one minus power. So it may be that we find no significant difference between dissimilar detectors, we keep using our old detector, miss detections, and potentially result in loss of life. Essentially we have missed a real effect. This error is harder to account for.

One of these errors will be more important than the other, in our detector example loss of life and therefore the power level was more important. An example of the inverse could be associated with improvised explosive devices (IEDs) where we could be investigating a new technique for rendering them safe. If we say the technique has no effect when it does in reality (power), then that isn't as important as saying the technique has an effect, and therefore declaring it safe, when in reality it doesn't (confidence), as that could result in loss of life if taken forward.

Numbers translated: 90% confidence means that there is a 10% risk, or a 1 in 10 chance, that we could find a significant difference when we shouldn't have. 95% power means that there is a 5% risk, or a 1 in 20 chance, that we may miss a significant difference when there should have been one.

Ideally we want to minimize the risk of these errors. However by doing so we increase the sample size needed, so it's a balance between an acceptable level of risk and an affordable sample size. It's quite complicated, but if you can translate confidence and power to your customer in terms of risk and what that would mean in the study you're working on, it should make it clear so you can get the values, or range of values, you need.

Continuous Data

When dealing with continuous data you need three of the following four: a confidence level, a power level, an effect size, and a sample size. The fourth is only needed if your sample size is limited.

The effect size is calculated from a meaningful difference and an estimate of the standard deviation. It is the difference divided by standard deviation, and this is then used in the calculation.

A meaningful difference is a difference that is of scientific interest and is quoted in the units that will be measured. It is also known as the delta. Using our marathon runners' example, a meaningful difference may be 60 seconds.

An estimate of the standard deviation can be tricky. The first step is to determine whether there is any historical data that can be used. If there is, problem solved, otherwise you may need to look down other avenues. One is through literature, if there are relevant examples available. Another option is to first run a pilot study, and then run a power calculation using the standard deviation from that data to decide how many more samples need to be taken to satisfy the required power or confidence levels.

If you are constrained by resources you may already have a set sample size, in which case you would be running the calculation to find the effect size, the power, or the confidence level to determine whether it has achieved a satisfactory value.

Binary Data

If you are dealing with binary data you need three of the following four: a confidence level, a power level, an effect size, and a sample size. Again you need the fourth only if the sample size is limited.

The effect size here is just a difference in proportions, e.g., a difference of 0.1, which is equivalent to 10%. However due to the binomial distribution being bound by 0 and 1, you need to specify where that proportion will be and this is what is known as *worst case proportions*. For example, if you know what you are testing, let's use detectors, will be at the high end, say around 90% and you were interested in a 10% difference, then you could chose a worst case proportion of (0.85, 0.95). However if you have no idea about the performance of the detectors then you would need to choose (0.45, 0.55), which straddles the 50% mark. The sample size will be larger using the middle values as opposed to either end due to the fact that the values are not near the bounds, but this is required if there is no previous knowledge.

An effect size is then calculated from these worst case proportions by the software and is used in the calculation.

Conducting the Calculations

The main package I use in R to run these calculations is called **pwr**. Within the **pwr** package there are multiple commands for the different data types and also test types. The structure for a continuous data calculation is as follows:

pwr.t.test(n = , d = , sig.level = power = , type = , alternative =).

Broken down, this is what each section represents:

- **pwr.t.test** is the command for running a power or sample size calculation for testing means of continuous data

- **n** is the sample size

- **d** is the effect size

- **sig.level** is the significance level, one minus the confidence level

- **power** is the power level

- **type** stands for which type of test you will be using in the analysis; options include: **"one.sample"**, **"two.sample"**, and **"paired"**; the default is **"two.sample"**; more on type below.

- **alternative** stands for what your alternative hypothesis will be and the options include: **"two.sided"**, **"less"**, and **"greater"**; the default is **"two.sided"**.

Type means whether you will have one sample of data, one set of numbers that you want to compare against a single figure; two samples, two groups that you want to compare against each other; or paired samples, two measurements per participant.

Example 1.1 shows how to run and interpret a sample size calculation for continuous data.

EXAMPLE 1.1

We are looking for a sample size calculation for our experiment involving two groups who throw a javelin. We just want to look for a difference between them. We have data from a previous experiment that suggests a standard deviation of 0.81 meters, and we are interested in a meaningful difference of 1 meter. Thinking about the risks, we are happy to take a 5% chance of seeing a spurious difference and a 10% chance of missing a real effect.

```
# Load library
library(pwr)

# Run sample size calculation
pwr.t.test(n = NULL, d = (1/0.81), sig.level = 0.05, power = 0.90,
    type = "two.sample", alternative = "two.sided")

Two-sample t test power calculation

n = 14.81761
d = 1.234568
sig.level = 0.05
power = 0.9

alternative = two.sided

NOTE: n is number in *each* group
```

Here we can see that each group needs to do 15 repeats (always round up) to achieve the specified confidence and power levels. So you can say 15 repeats are required in each group to achieve a confidence of 95% and a power of 90%.

Fifteen repeats is quite a reasonable number to complete. However if the sample size was much larger you may need to adjust the calculation by reducing the confidence or power levels, or alternatively increasing the effect size.

The structure for a binary data calculation is as follows:

pwr.p.test(n = , h = , sig.level = power = , alternative =).

Broken down, this is what each section represents:

- ***pwr.p.test*** is the command for running a one proportion power or sample size calculation for testing proportions of binary data; for two proportions the command is ***pwr.2p.test()***.

- ***n*** is the sample size

- ***h*** is the effect size; with the command being ***h = ES.h(wcp1, wcp2)*** where wcp is the worst case proportion 1 and 2

- ***sig.level*** is the significance level; one minus the confidence level

- ***power*** is the power level

- ***alternative*** stands for what your alternative hypothesis will be and the options are the same as with the continuous calculation

Currently there is no code to run a power or sample size calculation for paired proportions. However using the ***pwr.2p.test*** will be conservative and therefore can be used as a substitute.

Example 1.2 shows how to run and interpret a power calculation for binary data.

EXAMPLE 1.2

We are looking for a power calculation for our experiment involving two new detectors and we want to see if either one is better than the other. We can only run 10 repeats on each detector, but we have an idea of the performance through literature that states detection rates around 85%. We are interested in establishing a 10% difference between the detection rates, but we only want to take a 10% chance of seeing a spurious difference.

```
# Load library
library(pwr)
```

```
# Run power calculation
pwr.2p.test(n = 10, h = ES.h(0.80, 0.90), sig.level = 0.10,
    power = NULL, alternative = "two.sided")
```

Difference of proportion power calculation for binomial distribution

```
h = 0.2837941
n = 10
sig.level = 0.1
power = 0.1675034
alternative = two.sided
```

NOTE: same sample sizes

Here we can see the experiment would only achieve a power less than 17%, which means there's an 83% chance of missing a real effect. It is clear that this would not be acceptable so further discussion with the customer would be required to determine whether the sample size could be increased, otherwise it wouldn't be worth running the experiment as the results could be misleading.

In general a larger sample size will be required for binary data compared to continuous data. This is due to the fact that with binary data there are only two possible outcome responses, whereas for continuous data there is finer granularity. You can always convert continuous data into binary data for analysis if required, however you cannot do the inverse.

There are also options for power and sample size calculations for ANOVA, general linear models, chi-square tests, and so forth. However, these would require much more details and explanations so that can be left for your own further investigation.

Defining the Scope of the Study

All experiments are designed to have precision, coverage, and validity; generally there will be a balance between precision and coverage due to resource constraints. Through this there will also be assumptions that need to be stated.

Applicability of Results

It is very easy to start covering a wide range of scenarios, increasing the number of variables measured along with the number of repeats that would need to be run to cover the study space. It may be necessary to reduce the scope of the study to increase the precision concerning the reduced number of scenarios.

It is important to discuss the balance with the customer to determine whether being highly precise about a specific scenario is adequate to answer the study question. On the other hand, you may need to reach a certain level of precision that will put a limit on the breadth of coverage.

When determining the applicability of results some of the items that need to be considered could be environmental conditions or multiple operators. For example, if a piece of kit is being tested in cold and wet conditions, those results may not be able to be transposed to hot and sunny conditions. It may be that in your study that detail isn't important as the use will always be under the same conditions, but if not this needs to be accounted for. If there can be multiple operators it is not usually recommended to reduce the scope of the study by reducing it to one operator, there will be natural variation between operators and this needs to be recorded and taken into consideration when stating the results.

Assumptions

All assumptions need to be stated and recorded. These include assumptions about reducing the scope of the study, the variables, and the data. Assumptions about representativeness are very common: the test equipment, the operators, and the environment. There can be assumptions that the range within the variables chosen are adequate to cover the question space; that the variation between operators will be less than that of the natural variation in the system; that some methods used, such as randomization, will reduce the risk of bias in the results; and more.

Experimental Design

The design of experiments is planned such that the problem is investigated systematically to ensure sufficient relevant data is collected so that parameters can be estimated, the importance of variables can be tested, and consequently the study question can be effectively answered.

A poor experimental design, or a lack of one, can mean that any results need to be heavily caveated and may be misleading. In most cases a good experimental design can be created in a couple of hours, so even if there is a "fastball," there is still time to plan a design.

Variables

Choosing the appropriate variables and their ranges is linked with defining the scope, as it is the variables that are removed and/or reduced as necessary to reduce the scope of the study. There are multiple variables to think about:

- *response variable:* this is the metric for assessing the study question, there may be multiple response variables. It also can be referred to as the *outcome* or the *dependent variable.*

- *explanatory variable:* this is the measurement that may cause variations in the response variable, generally there are multiple explanatory variables. It also can be referred to as a *factor* and sometimes an *independent variable.*

- *nuisance factors:* these are things that can cause variation in the response variable, but they are not of interest themselves, such as day, operator, and so forth. These should be recorded for investigation in the analysis.

- *sources of variation:* these are things that also can cause variation in the response variable but cannot be controlled, such as the manufacturing process, temperature (sometimes), and so forth.

Interactions

When you are investigating explanatory variables against the outcome and are only concerned with the independent effect of each, this is termed as *main effects.* Figure 1-3 shows an example where only main effects are having an impact on the response.

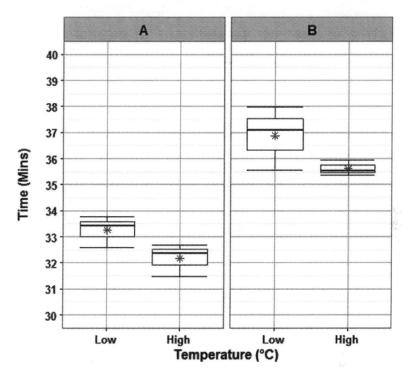

Figure 1-3. Example of main effects only

In Figure 1-3, Material A gives faster Time than Material B regardless of Temperature, and High Temperature gives faster Time than Low Temperature regardless of Material, although this needs to be tested not just plotted.

In some cases there may be interactions between the factors and the experimental design needs to be planned to account for this possibility. An interaction means two or more factors have a joint effect on the outcome. Figure 1-4 shows two examples of interactions.

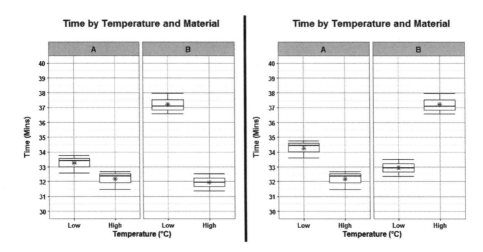

Figure 1-4. Example of interactions

The first plot in Figure 1-4 highlights that the interaction between Temperature and Material has an effect on Time due to the gradient slope. High Temperature gives faster Time than Low Temperature regardless of Material; however the slope of the gradient for Material B is much steeper, changing the Temperature has more of an effect on Time for Material B than Material A.

The other example of an interaction is due to the gradient changing direction. High Temperature gives faster Time than Low Temperature for Material A, but Low Temperature gives faster Time than High Temperature for Material B. Both of these interactions may have been missed if the experimental design wasn't done properly to include the possibility of interactions.

Confounding

If an experimental design doesn't account for these possible interactions then there may be confounding due to the fact you won't be able to tell which variable is having the effect on the response. For example, testing two cars on separate days, one car on each day, and finding a significant difference in the analysis. You wouldn't be able to confidently state whether the difference found was due to the cars or the days.

Example 1.3 shows a design that leaves itself open to possible confounding results, then a design using the same number of repeats that would allow the full testing of interactions.

EXAMPLE 1.3

There is a fastball that is investigating the accuracy of a weapons system. There are two factors that may have an effect on the probability of hitting a target. The two factors, height projectile is fired and angle of firing, both have four levels and there was money to fire 420 projectiles. Figure 1-5 shows a poor experimental design.

Weapon System Experimental Design		Angle (°)			
		15°	30°	45°	60°
Height from Ground (meters)	0m	60		60	
	1m	60	60		
	2m	60			60
	3m	60			

Figure 1-5. Poor experimental design

Although 60 repeats will give good precision, there is bad coverage due to the gaps in the design space. An analysis was run on the results and it was found that 1m gave significantly better probability of hit and 45° gave significantly better probability of hit. However we have no information about the relationship between the two factors. Due to this we don't really know which factor has the effect on the probability of hit: It may be one or it may be both.

Considering we have 420 projectiles we can run a better design while still obtaining good precision. Figure 1-6 shows a better design that would allow for testing the interaction as well as the main effects.

Weapon System Experimental Design		Angle (°)			
		15°	30°	45°	60°
Height from Ground (meters)	0m	26	26	26	26
	1m	26	26	26	26
	2m	26	26	26	26
	3m	26	26	26	26

Figure 1-6. Good experimental design

There are 26 repeats at each level that will still give good precision, and even leaves four tests left over in case anything goes wrong.

This time let's say that conducting the analysis on these results shows that the interaction was significant, which would have been missed using the last design. It was shown that if you use 1m then 15° was the best angle whereas if you use 3m then 45° was the best angle. This information would have been lost in the previous analysis and incorrect results would have been quoted to the customer.

It's not always feasible to run repeats at each combination of all the variables, and this is when it may be acceptable to have some confounding in an experiment. However, there are designs that can be used to allow for confounding while still retaining as much of the information as possible, see the Factorial and Optimal Designs section.

Designed Experiments

One of the primary ways in which data is collected is through designed experiments. In addition there are two main types of experiments: These are physical experiments and computer (or synthetic) experiments.

Figure 1-7 shows a quick comparison of physical and computer experiments.

Comparisons	Physical Experiments	Computer Experiments
No. of Factors	Relatively few factors (<10)	Many factors
Variation	Random variation - caused by environmental factors	Deterministic: No random error Stochastic: Randomness is modelled
Cost	Monetary cost	Time cost
Designs	Many standard design techniques	Deterministic: No standard designs Stochastic: Either standard or nonstandard designs - but use with caution

Figure 1-7. Comparison of various elements between physical experiments and computer experiments

Physical Experiments

There are a large number of design techniques for physical experiments, with the most common being factorial or optimal designs. I couldn't list all the available designs however I have mentioned a few of the other useful designs including adaptive designs in addition to the factorial and optimal designs.

To reduce the risk of bias it is good practice to use randomization where practically possible. Pure randomization will, as the name suggests, chose a random order from all the runs you need to complete. If that's not appropriate, then there is stratified randomization, which will still randomize the runs but will be bound by a factor.

Example 1.4 gives an example of how to use pure randomization and also stratified randomization.

EXAMPLE 1.4

We want to investigate two different paper test styles: differences in language, formatting, font, and so forth. There will be an equal split of Army and Navy personnel, each force type consisting of 20 participants. If we wanted to use pure randomization we would only use the ***sample()*** command in R.

```
# Create the list of test options
Test = rep(c("Test1", "Test2"), each = 20)
```

```
# Create a random sample list
Assignment = sample(Test, 40, replace = FALSE)
```

```
# Show the start of the random sample list
head(Assignment)
```

"Test2" "Test1" "Test2" "Test1" "Test2" "Test1"

Note that your output in both sections of this code will be different to mine as it involves randomization.

We could then assign each person walking through the door the next test on the list. However, this would be a case where we wouldn't want to use pure randomization as we may end up with all the Army participants sitting Test 1 and all the Navy participants sitting Test 2. We need to come up with a stratified random sampling list.

```
# Create a dataset of force type and participant
data = data.frame(Type = rep(c("Army", "Navy"), each = 20),
Participant = sample(1:40, 40))
```

```
# Split by force type then assign randomization to pick half of each
sp = split(seq_len(nrow(data)), data$Type)
samples = lapply(sp, sample, 10)
data = data[unlist(samples), ]
```

```
# Order list by participant number
data2 = data[order(data$Participant), ]
head(data2)
```

```
Type      Participant
Navy          1
Army          5
Navy          6
Army          7
Navy         11
Army         12
```

We now have a list of participants to assign to Test 1 and by default anyone else will be assigned to Test 2; and we will have an equal split of the forces between the two test types and therefore reduce bias.

Randomization is a good way of reducing bias and prevents some types of confounding that would impair analysis results. However it is not always practical and/or possible to completely randomize an experiment and this will need to be taken into consideration.

For example, you may be bound by a large piece of equipment that takes 20 minutes to set up each movement iteration. In these cases it would be more sensible to conduct several repeats while in one position. One way to counteract a possible bias could be to repeat the experiments again another day with the movement iterations in a different order. Otherwise the reason for not using randomization just needs to be stated in the technical report, such as the practicality of moving heavy equipment each time.

Factorial and Optimal Designs

The best design to be used in all cases, in an ideal world, is the full factorial design as this would provide you with all the information about the variables you have collected data for in addition to all their interactions. However, as the number of variables increases the required runs without any repetition increases immensely.

When all factors only have two levels the number of runs will be 2^f, where f is the number of factors. For example, with 2 factors the number of runs will be 4, for 4 factors it will be 16, for 8 factors it will be a huge 64, and so forth.

The full factorial design lets you completely cover the design space and look at all the interactions including the higher interactions; it also is balanced (see Example 1.5).

A fractional factorial design is based on a subset of the full factorial—such as half, quarter, and so forth. It allows for an investigation of all the main effects and some of the interactions between them. The advantage of the fractional factorial design is that it is balanced, which makes interpretation slightly easier.

An optimal design is also a subset of the full factorial but it's a noninteger fraction, such as 11 out of 16. It permits investigation of some or all of the main effects and some of the interactions, but there is more confounding. It is rarely balanced, which makes the analysis and interpretation slightly harder.

Example 1.5 uses the **AlgDesign** package.

EXAMPLE 1.5

We want to explore through-barrier detection where the aim is to see inside a container. We have decided to look at four factors each with two levels: the type of container (A, B), the chemical inside the container (C1, C2), possible interferents (I1, I2), and two concentrations (Low, High). We look at three possible cases, first if we could do the full factorial; second if we could only do half, fractional factorial; and last if we could only do eleven combinations, optimal design. In all examples I show a tidier example of the output data from R.

```
# Load library
library(AlgDesign)

# Create the full design and print the results
# This is also the full factorial design
des = gen.factorial(levels = 2, nVars = 4, center = FALSE,
varNames = c("Concentration", "Interferent", "Chemical",
"Container"))
des
```

Figure 1-8 shows the full factorial design with the factor levels translated from 1 and 2 (R output) to the levels described earlier. Figure 1-9 shows the breakdown of options to highlight the balance in this design.

Combination	Container	Chemical	Interferent	Concentration	Combination	Container	Chemical	Interferent	Concentration
1	A	C1	I1	Low	9	B	C1	I1	Low
2	A	C1	I1	High	10	B	C1	I1	High
3	A	C1	I2	Low	11	B	C1	I2	Low
4	A	C1	I2	High	12	B	C1	I2	High
5	A	C2	I1	Low	13	B	C2	I1	Low
6	A	C2	I1	High	14	B	C2	I1	High
7	A	C2	I2	Low	15	B	C2	I2	Low
8	A	C2	I2	High	16	B	C2	I2	High

Figure 1-8. Full factorial design

Full Factorial		Interferent 1		Interferent 2		Total
		Conc. Low	Conc. High	Conc. Low	Conc. High	
Container A	Chemical 1	1	1	1	1	4
	Chemical 2	1	1	1	1	4
Container B	Chemical 1	1	1	1	1	4
	Chemical 2	1	1	1	1	4
Total		4	4	4	4	16

Figure 1-9. Full factorial design balance

```
# Create the fractional factorial design - 8 out of 16 combinations
half.des = optFederov(data = des, nTrials = 8, approximate = FALSE)
half.des$design
```

Figure 1-10 shows the fractional factorial design with the factor levels translated from 1 and 2 (R output) to the levels described in this example. Figure 1-11 shows the breakdown of options to show the gaps but also to highlight the balance in this design.

Combination	Container	Chemical	Interferent	Concentration	Combination	Container	Chemical	Interferent	Concentration
1	A	C1	I1	Low					
					10	B	C1	I1	High
					11	B	C1	I2	Low
4	A	C1	I2	High					
					13	B	C2	I1	Low
6	A	C2	I1	High					
7	A	C2	I2	Low					
					16	B	C2	I2	High

Figure 1-10. Fractional factorial design

Fractional Factorial		Interferent 1		Interferent 2		Total
		Conc. Low	Conc. High	Conc. Low	Conc. High	
Container A	Chemical 1	1			1	2
	Chemical 2		1	1		2
Container B	Chemical 1		1	1		2
	Chemical 2	1			1	2
Total		2	2	2	2	8

Figure 1-11. Fractional factorial design balance

```
# Create the optimal design - 11 out of 16 combinations
op.des = optFederov(data = des, nTrials = 11, approximate = FALSE)
op.des$design
```

Figure 1-12 shows the optimal design with the factor levels translated from 1 and 2 (R output) to the levels described in this example. Figure 1-13 shows the breakdown of options to highlight both the gaps and the imbalance in this design.

Combination	Container	Chemical	Interferent	Concentration	Combination	Container	Chemical	Interferent	Concentration
					9	B	C1	I1	Low
2	A	C1	I1	High	10	B	C1	I1	High
3	A	C1	I2	Low	11	B	C1	I2	Low
4	A	C1	I2	High					
5	A	C2	I1	Low					
					14	B	C2	I1	High
					15	B	C2	I2	Low
8	A	C2	I2	High	16	B	C2	I2	High

Figure 1-12. Optimal design

Full Factorial		Interferent 1		Interferent 2		Total
		Conc. Low	Conc. High	Conc. Low	Conc. High	
Container A	Chemical 1		1	1	1	3
	Chemical 2	1			1	2
Container B	Chemical 1	1	1	1		3
	Chemical 2		1	1	1	3
Total		2	3	3	3	11

Figure 1-13. Optimal design balance

The first example shows the complete balance by running all the experiments, the second shows that there is still balance while containing gaps in the design, and the third shows how due to the difficult number of experiments balance can't be maintained, although the design does its best. The examples also show how simple it is to generate the designs in R.

You should remember that when using a fractional factorial or optimal to take care when interpreting some main effects or interactions that will be confounded with higher term interactions.

An example of acceptable confounding would be to confound a lower-order effect, such as an interaction between two factors, with a higher-order effect, such as an interaction between three factors, as it can be hard to interpret the higher-order interactions. Here we make the assumption that the observed differences are due to the lower-order effect rather than the higher-order effect.

Adaptive Designs

There may be some cases where an adaptive design is required, and this just means exactly that, the design is adapted during the experiment due to the results found.

Some examples of this are when you look for limits of detection for a new piece of equipment, survivability curves, or protection rates at different threat levels.

There are some common methods used when dealing with binary response data in terms of a continuous explanatory variable, one of which is the Langlie method (see Figure 1-14).

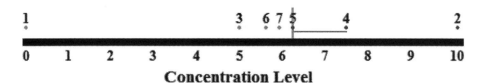

Concentration Level

Figure 1-14. Example of the Langlie method

Based on subject matter expert (SME) judgment a "window" of values is chosen where you are sure that the lower end will definitely produce negative results and the upper end will definitely produce positive results or vice versa depending on the scenario.

Here the first experiment is run at almost the 0 concentration level, which gave a negative response. Then the second experiment is run at the 10 concentration level, which gave a positive response. The third experiment is run at half the concentration between the two, and as this produces a negative response the fourth experiment is run at halfway between the concentration of Experiment 2 and Experiment 3, as we know we need to increase the concentration for a positive result. This continues until you have reached a concentration level you are happy with, or until your repeats run out. This will however only let you know the 50% detection limit.

The issue with this method is there is no correction for either false positives or false negatives. For example, in Figure 1-14 let's say that Experiment 5 was a false positive, for the rest of the trial all experiments will be conducted under that concentration level; however the actual area of interest lies between the concentration levels of Experiments 4 and 5, which now will never be investigated.

Another common method is the Staircase method, see Figure 1-15.

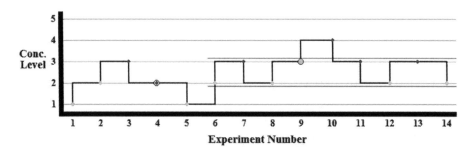

Figure 1-15. An example of the Staircase method

Based on SME judgement both a "window" of values and a step size is chosen so you not only have each "end" of the trial, but also set steps between those values. It can be tricky to pick a sensible step size as you need it to be big enough that you reach the area of interest without lots of repeats, but you also need it small enough that it's a sensible set of values.

Here you can start at either end. We started at the lower end, and then ran the first experiment. This was a non-detect so Experiment 2 was run at the next step up in concentration. This continues going up or down steps depending on whether the response was detect or non-detect.

A positive thing about this method is that it can account for false positives and false negatives, as circled in Figure 1-15, as the next time that step is reached there is a chance of the correct result. At some point the area of interest will be reached and you will "bounce" between the steps, so you will know you are in the correct location, as shown by the dark grey lines. The downside of this is the step size defines the granularity you can see.

This leads on to adaptive designs, the Staircase method is a good way of narrowing down the area of interest with a small sample size. You can then take that step, concentration level 2 and 3 in Figure 1-15, and divide that up into sections for further testing, which is the adaptive part of the design.

This can be split into as many sections as you wish, but you should include some repeats at each section. You should also include a section within the 20% to 30% area and within the 70% to 80% area to obtain the most information as a curve will be fitted to the results, see Figure 1-16.

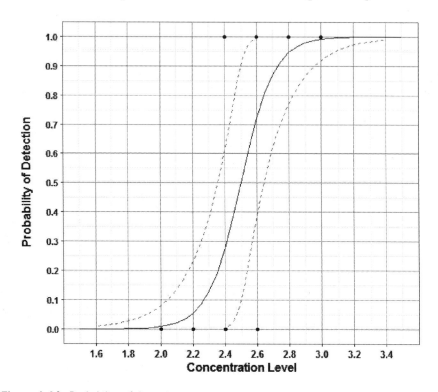

Figure I-16. Probability of detection curve

From Figure I-16 you can now find any other probability of detection, not just the 50% mark. You also can add different level confidence intervals to the curve.

Other Designs

There are many other designs; however I will only mention a few key designs used for different types of experiments.

Screening designs are used when a large number of factors need to be explored using a small number of experiments, such as only being able to use 12 experiments for 11 factors each with 2 levels. A common type of screening design is the Plackett–Burman. These designs are useful only if the main effects are important and not the interactions as the main effects will be confounded with the two-way interactions.

Randomized control trials are used most frequently in medical research and participants are randomly assigned to groups. The objective of these trials is to compare treatments to assess the efficacy of interventions where participants can only take one treatment. An alternative where it's appropriate to take multiple treatments is called a cross-over design. However, the process of randomization of participants can be quite complex.

Mixture experiments are used when the volume/mass is fixed and the investigation is for the proportions of the components. The constraint with this approach is that the level of one component is automatically defined once the other component levels are fixed. The design space can get quite complex with multiple factors and if there are constraints on the mixture itself. These designs are very common in the food sciences.

Computer Experiments

There are no standard approaches for deterministic computer experiments as there is no variation in the results, if you ran the model 20 times with the same factor set up you would get exactly the same answer each time.

Stochastic computer experiments have the variation modeled in, a lot of the time this will have been through a mix of physical trials and/or SME judgment. Again, as with the previous section I only discuss a few standard designs in addition to physical design methods.

Latin hypercube sample (LHS) is a commonly used design. The factors are scaled from 0 to 1 and are scaled based on the minimum and maximum values for each one. If the computer experiment has time to run to completion this is a recommended method as it will thoroughly cover the system space. However if the experiment has to be stopped early then there may be spaces with no information due to the randomization of placing the points.

Low discrepancy sequence also has its factors scaled from 0 to 1 by the minimum and maximum values. However the advantage of this over the LHS is if the simulation may be stopped early as this method pushes the points around the design space. The algorithm works by trying to fill the empty spaces so if it does have to finish early there is still a good spread of information around the design space. The disadvantage though is that it takes longer to run than the LHS. Figure 1-17 shows an example of the difference between the LHS and the low discrepancy sequence if stopped after 10 runs instead of 25 runs.

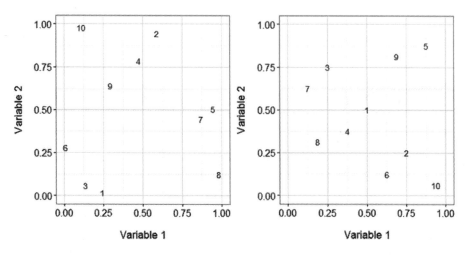

Figure 1-17. LHS and low discrepancy sequence designs stopped after 10 runs

Uniform designs are useful if you are only interested in an average, which may not be ideal if the response is complex and has regions of extreme change. This design is another that scales the factors from 0 to 1, but it then puts points on a neat grid rather than randomly scattered through the space.

An example where a uniform design wouldn't be appropriate is using a simulated motorcycle accident, as data shown in Figure 1-18.

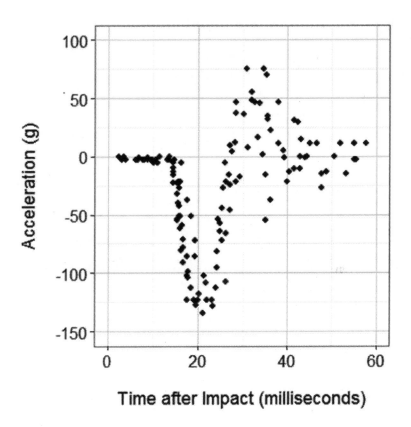

Figure 1-18. Relationship between variables from a motorcycle accident

The measurements recorded are head acceleration across time after impact. It is clear that this is a complex relationship as the head jerks back and forth in an accident. Figure 1-19 shows the difference between using a uniform design compared to a LHS design for investigating the relationship between the variables.

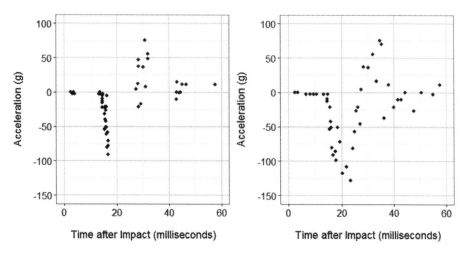

Figure 1-19. Uniform design compared to a LHS design to model a motorcycle accident

If the uniform design had been used a lot of important information would be missing and we wouldn't have seen the pattern discovered through using the LHS design.

When there are very complex models that take days to run iterations of the simulations, meta-models can be developed to give a quick response. A sample will be required to train the meta-model, and it should include uncertainty about the predictions. Once trained it should be a good representation of the complex model that will produce much quicker results. The tradeoff is that the response will be less precise, but it's a good indicator before the complex model results are completed.

At the observation points of the meta-model there is little or no variation, however between the points the variance increases, and once beyond the limits of the outer points the variance increases exponentially, so care should be taken with the estimates made outside the recorded range.

Surveys

The other primary way in which data is collected is through surveys. Some of the advice also can be used in questionnaires given to participants during a designed experiment. However survey design as a data collection method generally refers to the case where the participants are not known or are assigned beforehand.

Figure 1-20 shows the suggested process to follow when thinking about designing any survey.

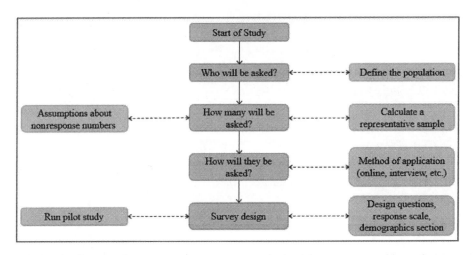

Figure 1-20. Survey design thought process

Once you have assessed who the population includes and the population size, a representative sample size will need to be calculated. However this will need to be tweaked due to nonresponses. Response rates for surveys can be quite low, so you will need to send out extra surveys to account for this. Additional thought should be given to the reasons for nonresponses and how this may cause bias in the results and affect the scope of the study.

When designing the survey itself and this applies to designed experiments questionnaires as well, there is quite a lot to think about. Personally I consult a human factors specialist as it's not my area of expertise, but the following are general areas for consideration.

Wording of the questions can be quite tricky, especially if a suitable survey design doesn't already exist. Keep the language simple but precise as to what you are asking, keep the questions short, avoid leading questions, determine whether the question is multiple choice, and don't always be biased toward using positive statements.

Think about the question response type and the implications during analysis they may have. For example, if an open text box was the only response option available, which analysis methods could be used? However, it can be useful to provide an optional text box as it can help clarify why participants have selected a specific answer. Likert responses are a popular method for questionnaires, but thought needs to be given to whether there is an even or odd number of categories, that is, do you allow an "on the fence" option. There are multiple opinions about this, but it's ultimately a choice whether you want to force a subject to pick a positive or negative answer when they may not want to or allow a neutral answer that may not be informative in the analysis.

Demographics also should be collected whether it's a key focus of the survey or not, as they can help inform results during the analysis. Again there is discussion as to whether this should be placed at the beginning or the end of the questionnaire.

It's always recommended to run a pilot study first as this will highlight any ambiguity in the questions, suggest where closed questions should include more response options, show where a question would be better placed in a different format, and also indicate whether the survey is a manageable length.

Once the experimental design has been completed, the next item to do is think about is collecting the data and formatting it.

Summary

This chapter delved into the thought that needs to go into designing experiments and showed a suggested process to follow to make sure all the relevant information is collected. It was split into four main sections corresponding to the four steps shown in the design of experiment process in Figure 1-1.

The first of which showed how to form the study question that will then be turned into a hypothesis. It explained briefly how hypotheses are formed and also described the type of questions that should be asked of the customer.

The second section looked at power and sample size calculations and showed some common misunderstandings involved with the "how many" question. Within this section the first subsection explained the information needed to run the calculations, such as risk and other values dependent on data type, and how to translate that to get it from the customer. The second subsection then delved into how to actually conduct the calculations using R and how to interpret the results.

The third section discussed defining the scope of the study including thinking about the applicability of results in other scenarios and also the assumptions that need to be justified and recorded.

The last section was the largest and moved onto how to design the experiments themselves. This was split into three subsections, variables, designed experiments, and surveys.

The first subsection described some common terminology linked to variables, and then went on to explain interactions and confounding effects with examples.

The second subsection probed into some of the available designs for both physical and computer experiments. This included factorial and optimal designs, adaptive designs, and some of the other popular designs for physical experiments, as well as some off the common computer experiments designs including Latin hypercube samples, low discrepancy sequences, and uniform designs.

The final subsection briefly investigated survey design including a suggested process of what needs to be thought about along with some further details of each stage.

Chapter 2 moves to actually collecting the data and what considerations should be given before jumping straight into carrying out the experimental design created. It also looks at the formatting side of things once the data is being/has been collected.

Data Collection

How Do I Get the Data?

The main aim of writing an experimental design is to make sure you collect the relevant data to answer the customer's question. The next step is to create a data collection plan to ensure that the actual data is both available as soon as possible and also in a useable format.

Collecting the Data

When collecting data the most important thing is to create a data collection plan. This informs everyone involved as to what data needs to be collected, how it is to be collected, when it needs to be collected, and why it is being collected (i.e., which key question that particular data will help to answer).

Along with this plan there should be clear instructions to the observers, where applicable, who will be gathering the data, and this information will vary dependent on the type of data that is being collected; mainly split into objective and subjective data.

Figure 2-1 shows the suggested process to follow when thinking about all the aspects involved in data collection.

©Victoria Cox 2017
V. Cox, *Translating Statistics to Make Decisions*, DOI 10.1007/978-1-4842-2256-0_2

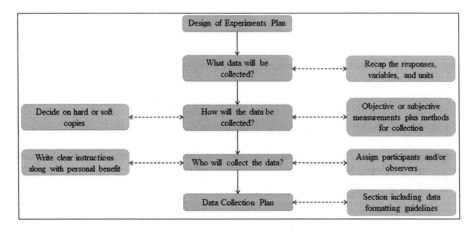

Figure 2-1. Data collection thought process

More detail will be given during the later sections of the chapter. However a brief description about each section of the process as shown in Figure 2-1 is given here.

The first section of the data collection process is actually a repeat of the final section in the design of the experiment's plan, as thinking about the data collection plan stems from the design of experiments plan. This will just get you rethinking about the different variables and what you will actually be collecting along with the units in which these measurements will be taken.

Second is to think about how the data will be collected. Therefore by initially dividing the variables into whether they are objective data or subjective data will affect the following questions. Methods of collection need to be decided, and these will generally vary dependent on the answer to the previous partitioning; for example, questionnaires will mainly be used to collect subjective data. The final slightly disjointed item to be considered is concerning the collection of the data and whether the collected data will be initially recorded on hard or soft copies, or a mix of both.

The next section involves considering who will be collecting the data, again this may vary dependent on the data types chosen earlier. Whether you just need to assign participants, recorders of objective data, or observers, they all need clear guidance as to what is expected. In addition to this, adding the personal benefits to participants you have chosen to partake in the trial will encourage full thoughtful answers.

Finally all of this needs to be recorded in a data collection plan for reference. There also should be a section within this plan for the data formatting guidelines, see the Formatting section.

Objective Data

Objective data collection generally applies to observers recording figures according to machine or tool readouts. It is slightly confusing as it also can apply to measurements not strictly taken using a tool, but more of a subjective statement of an objective measurement—see the example in the Variation section. In this section I discuss objective data as that collected during a physical experiment by recorders.

Variation

Dependent on the measurement type, the value stated can vary between recorders due to natural human variation even during a physical experiment.

For example, if you are asked to note the time when liquid has started to absorb through a material and changes the color of the material, your answer may not be exactly the same as someone else's looking at the same test. This is why multiple observers are recommended even with objective data; unless it is a machine readout that states the figure and just needs copying. Although the absorption time is, strictly speaking, a subjective measurement, it is defined more as an objective measurement due to the fact that a machine could record it, if one existed. It is not truly subjective, the variation between observers will be slight and due to eyesight or something similar rather than personal opinion.

Tied in with using machines is the fact that there most likely will be machine variation that will need to be accounted for as with human variation. Both of these things need to be considered during the design of experiments phase. Where appropriate, consideration should be given to calibration of the machines: is it necessary to ensure multiple machines have been calibrated at similar times, have they been recently calibrated or was it a long time ago, also what if one machine needs to be calibrated half way through the experiment? This is a data collection issue as the design of experiments will have only covered the fact that you need multiple machines to cover process variation.

Repeats

It is good practice to take multiple measurements within recorder tests as this also can have some variation. For example, when testing ammunition one of the tests looks at the dimensions and another test looks at the weight. For the first test the recorder is measuring the ammunition with a ruler and is recording the values, for the second test the recorder is using a set of scales. The same answer may not be achieved for every repeat of the same ammunition and therefore a few repeats should be taken to account for both the recorder and machine (scales) variation. In terms of extra data for analysis it will add no time at all, and in terms of recording the extra measurements it will add on very little time and effort to the experiment; plus it will also give a much better estimate of the real result.

Precision

Additional information should be given about the precision requirements for the measurements taken. For example, should the recorders note the result to one decimal place, two, three, and so forth? This is important to state upfront as otherwise you may get one recorder that rounds the figures to the nearest whole number while your experiment requires specific detail about the equipment performance.

Frequency of measurements also needs to be stated before the experiment begins. For example, if there were a set of runners that were recording their heartrate over a set period of time they need to know how often they should record it so the measurements are consistent across the runners and therefore will be comparable in the analysis. In this instance each runner should have the same type of equipment, which would have been covered in the design of experiments. However, each piece of equipment should be tested on a single participant beforehand to establish an initial idea of the variation in the equipment; this is so the equipment variation can be accounted for during the analysis when human variation comes into play, and therefore won't be confounded with that variation. There should also be a record about which piece of equipment has been assigned to which participant.

How to record missing data or erroneous data, such as a value too low for the machine to detect, should be identified before the experiment begins to reduce the data formatting time. For example, Figure 2-2 shows some possible responses for missing data I have come across when dealing with Ct values,[1] which should be a continuous value.

---

[1] A Ct is the number of cycles taken for a fluorescent signal to cross a threshold (cycle threshold); this is used to detect a positive reaction in a PCR (Polymerase chain reaction) assay.

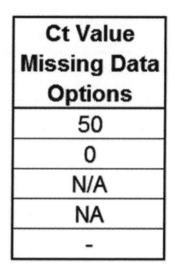

Figure 2-2. Missing data responses for Ct values

Subjective Data

Subjective data applies to both responses from participants about their opinions and also observer notes about participants they have been asked to observe.

Participants

When collecting data from participants, this is primarily done by questionnaires. For this section we focus on questionnaires that are physically filled out by the participants themselves.

Instructions

The key thing is to give instructions to the participants not only about what is required from them, but also about why the information is needed. This may seem like an obvious statement but the second point isn't always applied.

There's no better incentive than one that shows exactly how the responses collected can directly benefit the participant. For example, working with military participants, they aren't too enthralled about filling in multiple questionnaires, but if the benefits are made clear, such as this will directly aid with which body armor you will be using in the future, then they are much more willing to put in the time and answer the questions thoughtfully. There's nothing worse than receiving blank questionnaires or questionnaires where no thought has gone into it. For example checking exactly the same response throughout or always checking the neutral response due to them being either too busy and/or not engaged.

Giving clear instructions about how to fill in the questionnaires properly will reduce the data formatting time immensely. Even with a well-planned and produced questionnaire, participants can still answer incorrectly, in terms of the response data type that is required for the analysis. For example, if you are using a Likert-type scale such as in Figure 2-3 for the likelihood of an event occurring, definition needs to be given as to the responses you are expecting to see.

| Almost
Certain | Highly
Likely | Likely | Evens | Unlikely | Highly
Unlikely | Amost
Impossible |

Figure 2-3. Likelihood of an event ocurring response options

Figure 2-4 shows an example of some of the responses received that we would class as "incorrect" for use in the analysis due to there being multiple options chosen. "AC-AI" in the figure below stands for "Almost Certain–Almost Impossible" on the scale in Figure 2-3, and so forth.

Actual Responses			
AC-AI	L/E	HL/E	U-HU

Figure 2-4. Likelihood of an event ocurring responses

In the questionnaire provided, the question simply stated "what is the likelihood of the event occurring?" with the Likert scale in Figure 2-3 shown on its side at the edge of the page, and a blank line for the participants response. The first "incorrect" response could clearly be discounted from the analysis as the participant had covered the entire response option space however the other responses are understandable as they are within a close range of each other. A lesson learned from this would be to either specifically define up front that only one likelihood could be chosen or to offer boxes stating that to check one option only.

A recommended extra for questions similar to the one in the previous paragraph would be to include a confidence level. For example, in the questionnaire, below each question, we included a five-point Likert confidence scale from very confident to very unconfident. The advantage of this was that it specified their confidence in the answer they chose so it aimed to improve the understanding of the analysis. You could see when someone really wasn't sure about an answer, and may have just chosen it because they knew they needed to note an answer, and also when someone was very confident about the response they had chosen. This isn't always applicable though, such as for questions about comfort, ease of use, preference, and so forth.

Repeats

It may be necessary to gather multiple questionnaire responses from the same participants over time, in which case this needs to be accounted for during the analysis. It is during the design of experiments phase that the number of questionnaires per participant needs to be decided on dependent on the relative gain of useful information. However, it's worth noting that running a pilot study may improve the estimate.

Taking this approach, while beneficial to answer the customer question, it can be cumbersome to the participant and thought needs to be given to that aspect. Imagine that you were the participant having to answer all these questionnaires, could you give thoughtful answers each time?

We conduct multiple studies where we piggyback, as it were, onto military training exercises; so they are performing their everyday routines without any interference from us except for giving out the questionnaires. From our point of view we want to gather lots of information from them once we have added something extra to their training, such as different vehicles, storage, clothing, and so forth, and from their point of view they just want to get on with the training. We have to come up with a balance of gaining our information without annoying or distracting them which would result in noncooperation or unrealistic responses. An easy starting point, as I've mentioned earlier, is making sure they understand the direct benefit to themselves by giving us this information, then making sure the questionnaires are as clear and concise as possible, and finally having the potential to be a bit flexible. For example with the last point, you don't want to be stopping them mid-fire with the "enemy"; postponing the questionnaire until just after would produce much better results as they won't be rushing to get back to the action.

Observers

In some cases the subjective data will be collected by observers watching other people, the participants, carry out an experiment, or training using the military example. This also can be done in addition to the participants directly filling in questionnaires themselves.

Instructions

Clear instructions need to be given to the observers as well as to the participants, as the observers need to know how their responses will fit into answering the key customer questions. They also need to know exactly what the participants will be doing.

The observers need to be made aware of the type of things to look out for so that the notes they are making can be applied to one or more of the key question areas. For example, is it of interest for them to record who speaks to whom, the stress levels, if anyone is not being included, and so forth.

Tied in to this is defining who the observers will be watching, will there be different observers per participant or group of people, or will there be multiple observers generally watching the whole room?

If things such as stress levels are of interest should it just be recorded in note form or would it be preferable to record stress in a structured format, such as on a Likert scale, so there is comparison across both time and the observers? It's good practice to have the observers noting some of the same things that the participants themselves have been asked, such as work load, as it's interesting to compare the observers opinions to what the participants have actually recorded.

There is also the practical aspect of how long can the observer actually be alert to what is going on. There should be planned breaks for the observers with replacements coming in so the experiment doesn't need to be paused and the flow disrupted.

Authority

The observers need to know their place in a completely non-negative way. If their role is just to observe the room then they need to make sure they are as invisible as possible. They need to make sure they don't engage with the participants by asking or answering questions. The participants should be made aware that they are not to become involved with the observers. The observers also need to ensure that they stay impartial.

If the observer's role includes issuing and collecting questionnaires, or asking the participants the questions directly, then they need to the lead with this aspect. For example, there may be missing questionnaires in the results due to the observers feeling that as the military participant was a very high rank, they didn't feel they had authority to tell them what to do. It needs to be made clear that in an experiment they have the authority to stop the participant, in line with a sensible stopping time, to gain the responses required. The participant may get annoyed, but with reiteration of the personal gain from providing the responses, it should pave the way to get sensible responses rather than a rush through to get it done or a pure dismissal of the questions. This boils down to having good communication between the observers and the participants and it may mean changing observers around as sometimes there are personality clashes that just cannot be avoided.

Variation

There will be variation between the observed participants and that is what we want to collect through the experiment. However, there also will be variation between the observers, as there should always be multiple observers, and this we want to minimize as much as possible. A good starting point is to run a pilot study to highlight the type of items you want them to notice, collect, or do. This way they can practice as well as see the methods other observers use.

There are some statistics that can be used during analysis, or in the initial pilot, to measure inter-rater agreement and also internal consistency, two common statistics being Cohen's kappa and Cronbach's alpha. To be able to calculate these statistics an observer ID should be recorded.

Formatting

If the data collection instructions have not been very clear then the formatting of the data for analysis can take a long time, sometimes it can take longer than the analysis.

If the data has been collected by hand, that is, on paper; this will then need to be inputted into a computer. Thought should be given as to where these documents are stored to avoid being lost; how they can be distinguished from each other; a naming convention on the top is recommended; and when they should be inputted as soft copies, for example, at the end of each day, at the end of the experiment, and so forth.

Contrary to some advice given in statistics, when entering the data from hard copy to soft copy, this is generally best done by one person. This way there will be a much lower risk of data duplication and of different naming conventions. There also will be a point of contact for questions. However, for quality assurance the data also can be entered independently by another person, depending on the time burden, then both data sets can be compared on the computer to check for possible errors through human input.

If data has already been collected in soft copies then the task is to merge all the data and/or to check for the same mistakes that could be made inputting the data from hard copies, which we will look at later. Make sure there are always back up files for all soft copy data, the worst thing would be to lose a whole trial worth of data.

Generally speaking, how software requires the data to be set up for analysis usually isn't pleasing to the eye, so sometimes there will be a "viewable" data set and an analysis data set. If you are using R, it requires data to be stacked as in Figure 2-5, with the middle section deleted for space saving; unless you are dealing with paired data, which is discussed in Chapter 3.

Participant	Group	Gender	Question	Response
1	A	M	1	5
2	A	F	1	4
3	A	F	1	5
4	A	M	1	3
1	A	M	2	1
2	A	F	2	1
3	A	F	2	1
4	A	M	2	2
⋮	⋮	⋮	⋮	⋮
13	D	M	9	3
14	D	M	9	2
15	D	F	9	4
16	D	F	9	3
13	D	M	10	4
14	D	M	10	5
15	D	F	10	4
16	D	F	10	4

Figure 2-5. Example of stacked data

It's always better to start with the complete data set like this as it's very simple to subset out sections, such as per question, rather than start with separate columns for questions and have to merge them later on.

Different software may require the data to be set out in different formats, so you need to be aware of what format is required as this will drive how you structure the soft copy data.

I also recommend using Excel to store the data in csv files as it's accessible by almost all companies and can be converted if they have an older version. It is easy to use and can be read in most software packages.

That said, no matter whether you are inputting the data from hard copies or are having to check already entered soft copy data for errors/discontinuity, there are certain things you should be aware of:

- Get to know the software you will be using for analysis and what structure the data will need to be in.

- Make sure the data template is set up first, which means make sure you have all your column headers in place so you know where all the data fits.

- Ensure the column headings are kept as simple and clear as possible as a lot of software will convert spaces and/or symbols to a full stop. This doesn't apply to group levels within a variable as spaces and symbols can be included within these.

- Keep track of the data so you can avoid multiple entries or missing out data.

- If applicable, make sure the participants have an ID number, whether the questionnaires are anonymous or not, you may need to refer to it later on.

- Ensure you enter all the raw data rather than averages, any summary statistics can be done later using the software.

- Use the same value for missing data, whether it's a blank space or specific text.

- If there are zeros in the recorded data, check whether they represent zeros or missing data and input appropriately.

- When repeating text, such as group names, copy and paste is the best to use, as this way you avoid misspelling words, inserting an extra space, and so forth.

- If you are dragging a cell down for copying in Excel and it contains a number, make sure it copied correctly: copy the cell as opposed to continuing the sequence.

- Save the document in a useable extension for the software, for example R can read .xlsx files using certain packages but it's much simpler and quicker to save a .csv file.

Once the data has successfully been inputted and checked you can still double check items within the software. The software will soon highlight an error if the data is in an unreadable format; if there are extra unnamed variables, usually if there is a space in one cell; or if the data is unbalanced when it shouldn't be, if you haven't filled in all the data for one variable.

The following example of formatting checks, including some errors that may be seen, are shown using R, so skip to the end of the chapter if you won't be using that software.

Reading in data to R will immediately show if the data set is complete.

```
data = read.csv("MyData.csv")
Error in file(file, "rt") : cannot open the connection
In addition: Warning message:
In file(file, "rt") :
  cannot open file 'MyData.csv': No such file or directory
```

Common solutions to this error may be that R is linking to a folder that doesn't contain the data in which case you need to change the working directory, or make sure the data set is saved using the same extension as R is asking for.

Once the data has been read, successfully check the class of your variables to verify they are in a format you expect, it may be that you need to tweak one. For example, if Day was recorded as 1, 2, 3, then R will assume it to be an integer, treated as continuous data in models, and you may want it to be a factor.

```
class(data$Gender); class(data$Day)
[1] "factor"
[1] "integer"
```

```
data$Day = factor(data$Day); class(data$Day)
[1] "factor"
```

You also can check the levels of factors to ensure there haven't been any spelling mistakes or missing/extra levels.

```
levels(data$Smoker); levels(data$Likert)
[1] "N" "y" "Y"
[1] "1"  "12" "2"  "3"  "4"  "5"
```

You clearly can see there has been a typo when entering data for whether someone smokes or not (Smoker), and there also has been an error entering the Likert responses. With Smoker it is obvious that the "y" should be changed to a "Y" to tie in with the capital "N." However with the Likert response you don't know whether the value of 12 should have been a 1 or a 2, so either the hard copies will need to be referred to or the data may have to be excluded if there's no way of verifying the correct answer.

With numerical data you can investigate the minimum to maximum values to ensure that these give the range you were expecting, if there is an anomaly it may have been missed if only the mean was investigated.

```
min(data$Age); max(data$Age)
[1] 0
[1] 220
```

```
mean(data$Age); median(data$Age)
```
[1] 59.5
[1] 42

A quick look at the mean shows a sensible age, though it is a bit high if you also checked the median. However by including the minimum and maximum values we can see that there have obviously been some errors made with the data input. The age of 0 is clearly incorrect, but does it mean N/A or has another number been missed before the zero? Likewise with 220, should it have been 22, 20, or neither? These would both need to be checked and changed if possible; otherwise they would need to be excluded from the analysis.

As a side note if *summary(data$Age)* was used this would show the minimum, first quartile, median, mean, third quartile, and maximum of the data, more in Chapter 4.

You will be able to see if all the data has been entered, especially if you have equal groups of participants.

```
length(data$Response)
```
[1] 16

```
xtabs( ~ data$ParticipantGroup)
```
data$ParticipantGroup
　　A　B　C　D
1　4　4　　4　3

Although the length is correct for the number of responses we should have, the groups are not correct. One participant is not in a group when they should have been in Group D. The most likely explanation for this is that the last label was left off the spreadsheet by accident and is easily rectified.

Summary

This chapter has shown the thought process behind collecting the data in terms of what will be collected, how it will be collected, who will be collecting it, and the possible problems that can occur within each if not given enough consideration. It also looked at the issues that can occur with data formatting.

The data collection section was split into objective data and subjective data, as the two have different concerns with information that needs to be provided to recorders, participants, and observers and what is detailed in the data collection plan. The first gave thought to variation, repeats, and precision and the latter split into participants and observers. Attention was given to instructions and repeats for the participants, and instructions, authority, and variation for the observers.

The second section of the chapter then delved into issues that can arise with data formatting in terms of either inputting data from hard copy to soft copy, or just checking the original soft copy data. These issues included missing data, incorrectly labeled data, and having the data in the wrong format for use with statistical software, with the focus here being the R software. It also included other important points to consider. Examples with R code were provided to check for data formatting errors using the R software, some errors have been highlighted with suggestions as to why these errors may have occurred.

Chapter 3 takes us to after the experiment has been run and the data collected in a useable format. It looks at the first step before any analysis is conducted, which is exploratory data analysis (EDA). EDA is concerned with data types: viewing the data visually, identifying suspect data points, and choosing which distribution the data follows.

Exploratory Data Analysis

What Data Do I Have?

The first thing to do with any data set is to get to know it. This is done not only to familiarize yourself with all the data you have collected, but also to reduce the workload during analysis. The initial data investigation has been termed *exploratory data analysis* or EDA and it primarily focuses on visually inspecting the data. The main aim of EDA is to understand what data you have, what possible trends there are, and therefore which statistical tests will be appropriate to use.

Figure 3-1 shows the suggested process to follow when conducting EDA.

© Victoria Cox 2017
V. Cox, *Translating Statistics to Make Decisions*, DOI 10.1007/978-1-4842-2256-0_3

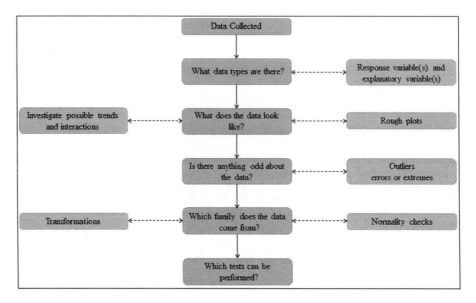

Figure 3-1. *Exploratory data analaysis (EDA) process*

The following sections delve into more detail for each of the steps shown in Figure 3-1. However the general idea is to identify the data types you have for each variable, for example, whether the data is continuous or discrete will lead to which plots can be created. These plots are initial investigations and Chapter 9 goes into more explanations about how to make clear, concise graphs to present to a customer as opposed to these quick and dirty inspections.

The next step is to identify any unusual data points and establish whether they are real outliers by using the plots. Once these have been dealt with the family of the data needs to be classified (e.g., normal distribution). By carrying out all these steps you can then move on to the final step that determines which tests can be performed on the data, but this is covered in other chapters.

Data Types

The type of data being collected should have been considered during both the design of experiments and data collection phase, however it is good practice to verify and if you haven't been involved from the beginning of the study it's a good place to start.

The main way to classify data types is into quantitative data or qualitative data; however the data also can be classed as univariate, bivariate, or multivariate. The latter three terms simply refer to the number of variables being recorded: uni (one), bi (two), multi (three or more).

In addition there are also the classifiers of objective or subjective data, which were mentioned in the previous chapters. The data being objective or subjective won't affect the tests used during analysis, however it will affect the assumptions and statements made in the conclusions.

Quantitative Data

Quantitative data is the term given to any data that is recorded as a numerical value. The subcategories within this are continuous data and discrete data.

Continuous Data

Continuous data is data that can be recorded as any value between an interval and as such it can be recorded with decimal points and still make sense (e.g., the strength of a signal in decibels or accuracy of a projectile in meters from a target).

Although age is usually recorded in integer form, it is generally considered to be continuous data due to the fact that you can be 21.5 years, but you just wouldn't record it as such.

Discrete Data

Discrete data is data that can only be recorded as an integer; it would not make sense to have 2.5 people for instance. Other examples include the number of canine detections or the number of survey responses.

Qualitative Data

Qualitative data is the term given to any data that is non-numerical and generally subjective. Qualitative data can be assigned a number to aid with analysis. However the precise value of the number itself is meaningless. The subcategories within this are binary data, nominal data, and ordinal data.

Binary Data

Binary data has two responses such as yes/no, heads/tails, and so forth (e.g., a detector detecting a target or the outcome of flipping a coin). When using binary data in analysis it is generally coded to 0 and 1 with 1 being the measurement of interest.

Binary data is actually a special type of nominal data, one with only two categories. When discussing the graph types later, any reference to nominal data will also include the case of binary data.

Nominal Data

Nominal data also is commonly referred to as categorical data, this is data that contains multiple groups that could be given a numerical value but have no natural ordering. For example, different types of vehicles such as bike, car, truck, boat, plane, could be assigned the numbers 1 to 5 for ease of analysis. However the values themselves are meaningless and have no ordering, car has the value 1 greater than bike by assignment only and not because it's "better."

Ordinal Data

On the other hand, ordinal data also can be given a numerical value but it does have a natural ordering. For example, reactions to a chemical could be none, rash, or blistering and they could be assigned the values 1 to 3 with 3 clearly being more severe than 1, but not necessarily 3 times more severe.

Another example of ordinal data is Likert responses from questionnaires, there is a clear progression from strongly disagree to strongly agree, or the equivalent. This data is not always treated as ordinal data as it should be, and as such the results can be misleading, see more in Chapter 7.

Viewing the Data

As mentioned earlier, in this section I discuss some of the different plot types that can be created given the types of data you have, although this list is by no mean exhaustive.

In EDA graphs are drawn to "get to know the data," so it's about noticing trends and structure for testing as opposed to drawing "pretty" plots. Chapter 9 is more focused on the effective presentation of graphs to highlight messages to the customer, including R code examples for amending color, labels, and so forth.

When plotting data regardless of the type of plot, the response variable should be on the y-axis, the vertical axis, and the explanatory variable should be on the x-axis, the horizontal axis. If there are multiple explanatory variables then this can be covered by different colors, shapes, or facets on the plot, for more information on graphs see Chapter 9.

Bar Charts

Bar charts are suitable for discrete data and counts of nominal data. There are many different variations such as stacked, percentage, and so forth, but they are very good to compare the frequency of different groups.

Figure 3-2 shows an example of the counts of different car makes bought at a local garage within one month. It clearly shows the order of car type sales: Ford, Vauxhall, Audi, Nissan, and then BMW.

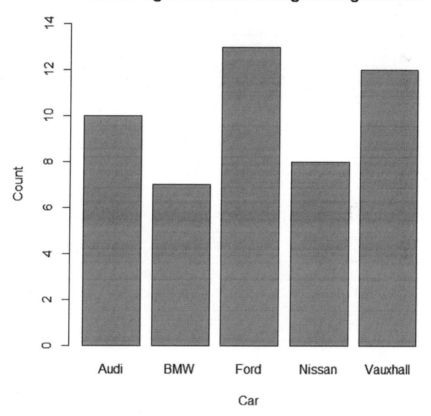

Figure 3-2. Bar chart of cars bought at a local garage

Dot Plots

Dot plots can be used in a similar manner to bar charts and are less cluttered as they show a single point as opposed to a bar. They can be used to show a single statistic, such as a mean, more clearly than a bar chart in which the filled bar below the top line of the bar would be redundant.

Figure 3-3 shows a dot plot of the average waiting times for baggage at selected United Kingdom airports, confidence intervals around the mean could be added to this plot for more information. At first glance the general trend is that the waiting times are shorter for Birmingham, East Midlands, and Manchester and are longer for Edinburgh, Gatwick, and Heathrow; which may be expected due to their size and popularity.

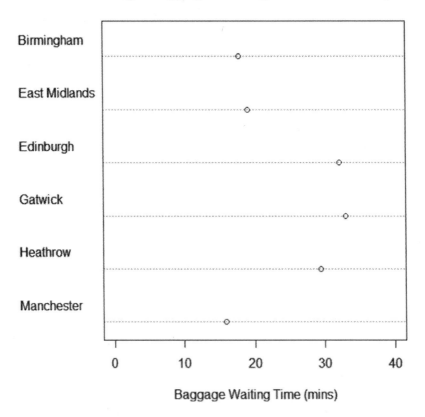

Figure 3-3. Dot plot of average waiting times for baggage at UK airports

Parallel Lines Plots

Parallel lines plots can be used to show paired data as the emphasis should be on the relative change for each subject. This information would be lost using any other plot listed. When creating a parallel lines plot the data shouldn't be stacked; there should be a separate column for subject, then two more columns for the items each person will be doing.

Figure 3-4 shows a parallel lines plot of the time to complete a task both before and after training for each subject. You can see that in all but one case the subjects completed the task quicker after receiving training.

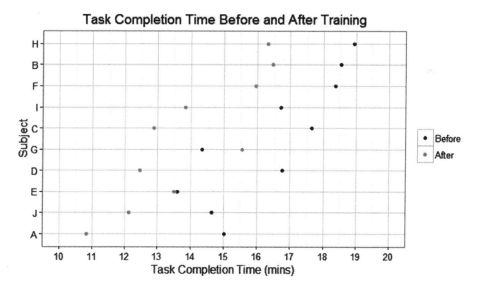

Figure 3-4. Parellel lines plot of training effect on task completion time

Histograms

Histograms are appropriate for continuous data only, they show the frequency counts given in set "bin" sizes. The software being used will choose appropriate bin sizes automatically. Histograms can be useful for highlighting distributions, such as a bell curve to suggest normality, however there is a better plot for investigating this assumption that is shown later in the chapter.

Figure 3-5 shows a histogram of heights from a sample of 100 people; in this case there is a bell curve that would suggest we could assume the data follows a normal distribution.

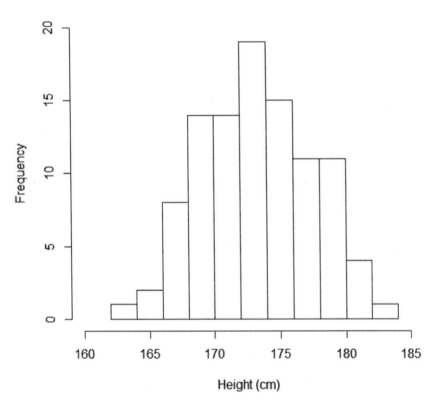

Figure 3-5. Histogram of heights sampled from 100 people

Scatter Plots

Scatter plots are useful for two continuous variables with or without a nominal data variable. These plots are handy for highlighting trends in the data as well as possible differences between any groups.

Figure 3-6 shows a scatter plot of yield by log concentration with a line of best fit. There is undoubtedly a positive trend between log concentration and yield and the points are quite close to the line of best fit. The line is plotted from a linear model, which is explained in Chapter 7.

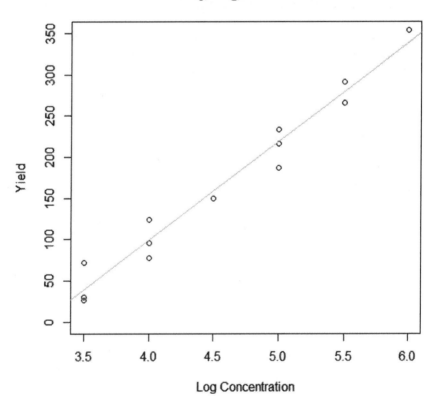

Figure 3-6. Scatter plot of yield by log concentration with line of best fit

Line Graphs

Line graphs are very similar in style to scatter plots. They are used for the same data types, but will generally have a time element across the x-axis. These points will be connected by a line to each individual point instead of a smooth trend.

Figure 3-7 shows a line graph of survey response rates by year; there was a sharp drop in 2011, but since then the trend seems to be picking up again.

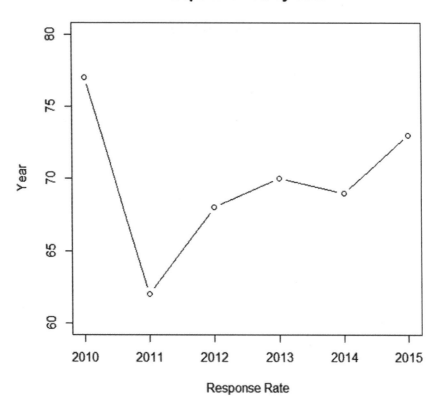

Figure 3-7. Line graph of survey response rates by year

Box Plots

Box plots are extremely useful as they can show a lot of information in a condensed manner. These plots are suitable for a continuous variable and a nominal variable. They can be used to investigate the distribution, equal sections suggest normality, however as earlier there is a better plot for this, and more important they can be very useful for highlighting differences between nominal groups.

A box plot also can be termed a box and whiskers plot as it principally concerns a box with some lines coming from each end. It contains the following information (descriptions of all of the summary statistics listed below are contained in Chapter 4):

- Median: the line within the box.
- Q1 and Q3[1]: the bottom line and top line of the box, respectively.
- IQR[2]: the length of the box itself.
- Range (minus statistical outliers): the length of the whiskers.
- Statistical outliers: any points outside the whiskers. The limits outside which a value is classed as an outlier are usually calculated as 1.5 times the IQR added to/subtracted from the quartiles.[3]

---

[1]Quartile 1 and Quartile 3
[2]Interquartile range
[3]Lower limit: Q1 − 1.5*IQR and upper limit: Q3 + 1.5*IQR

Figure 3-8 shows a boxplot of summer temperatures across the months June to August for popular holiday destinations. The larger the box and whiskers, the more variable the temperature, as for example Las Vegas; whereas the smaller the box and whiskers, the less variable the temperature, as for example London. In addition we also can start to see some possible differences we may find during testing, such as London and Paris having much colder temperatures than the other destinations and Florida and Las Vegas having much warmer temperatures.

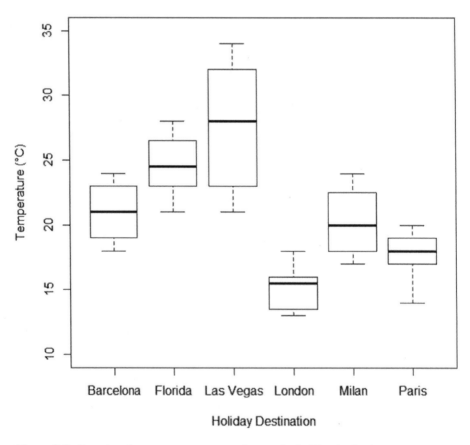

Figure 3-8. Box plot of summer temperatures by popular holiday destinations

Likert Plots

Likert plots are a clearer, more preferable way to view ordinal data rather than using bar charts. These plots are used to highlight the spread of data over the ordinal levels for different nominal groups. They can show the raw values or percentage values, which are preferable with unequal groups, along with the numerical count to the side.

Figure 3-9 shows a Likert plot of responses to three statements, "the equipment was easy to use," "the equipment was comfortable," and "the equipment was reliable over the week," for 15 participants.

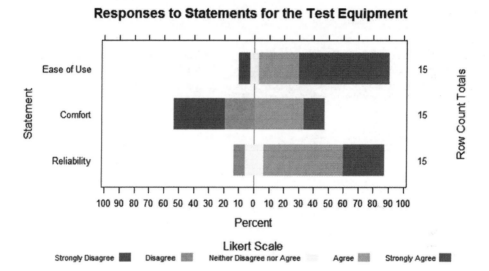

Figure 3-9. Likert plot of statement responses for the test equipment

The red and orange on the left represent the negative views, the yellow in the middle represents the neutral views, and the greens on the right represent the positive views. It can be seen that ease of use and reliability was rated more positive than negative. However, comfort was divided fairly equally between positive and negative.

Trellis Graphs

Trellis graphs are very useful for viewing multivariate data in a clear manner. This plot is limited to a maximum of two continuous variables, but can have multiple nominal variables. The limitation is dependent on the number of nominal variables given the number of levels within each variable. For example, if you had five nominal variables each with ten levels, it would not be sensible to plot them all on the trellis graph. A trellis graph creates multiple panels for nominal variables and also has the option to use shapes and colors for additional nominal variables.

Figure 3-10 shows a trellis plot of accuracy of a projectile by four explanatory variables: distance, ammunition type, operator, and target size. These four explanatory variables have 6, 2, 3, and 3 levels, respectively.

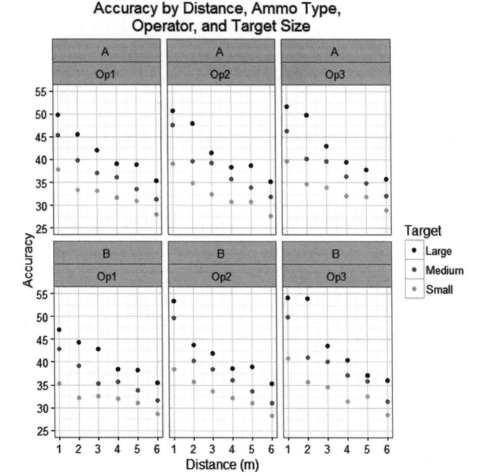

Figure 3-10. *Trellis graph of projectile accuracy by distance, ammo type, operator, and target size*

In all cases it seems that the accuracy decreased the further away from the target you got, which is probably to be expected. It also shows that the accuracy decreases as the size of the target decreases, again expected. There doesn't seem to be any major differences between the operators, comparing the left two boxes, to the middle two, to the right two boxes. There also doesn't seem to be any differences between the ammunition types, comparing the top three boxes to the bottom three boxes.

At initial glance it appears that there are no interactions of interest as there is no change of direction, such as the accuracy increasing as distance increases, in any of the boxes. There also is no change in gradient, such as the accuracy decreasing as distance decreases more rapidly, in any of the boxes; however, this should still be tested in the model.

These plots should be drawn to start to identify trends and group differences; however no conclusions should be drawn at this stage. Statistical testing needs to be undertaken to confirm the trends and add levels of uncertainty to the results due to the sample size and the data variation.

Outliers

There are two types of outliers, and each should be treated very differently otherwise any conclusions drawn from the data may be misleading. These can be classed as data entry errors/technical errors, or statistical outliers.

Data entry errors or technical errors are clearly incorrect data, such as 250% or someone who is 143. In both example cases the raw data would need to be checked to see if the true value could be obtained. However if there was no way of verifying this, then the values need to be removed from the dataset as they will heavily skew the data and influence the results.

Statistical outliers are those values that are highlighted as outliers through box plots and other similar graphs or through statistical testing for outliers such as using Grubbs' test. These outliers are possible, but are either at the extreme ends of the possible values or are just disjoint from the trend of the rest of the data. For example, someone with a height of 6'9" is a realistic value; however it will appear an outlier next to general height recordings. This should be double checked, and if the value was recorded correctly it should not be removed from the data set.

Drawing graphs can help highlight suspect data points that can then be investigated further to determine which type of outlier the points are.

For example, Figure 3-11 shows a scatter plot of the time taken to complete a race by participants of varying ages. It contains a data entry error that, if left in, would completely skew the results.

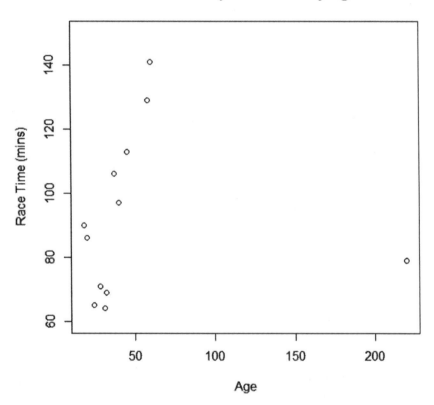

Figure 3-11. Scatter plot of time to complete a race by age, including a data entry error

Figure 3-12 shows the same dataset, but this time with the data entry error removed to highlight the difference that one data point would have made on the results. Now we can start to see a trend of race time decreasing with age until the "sweet spot" around the late twenties, and then see the race time increasing with age.

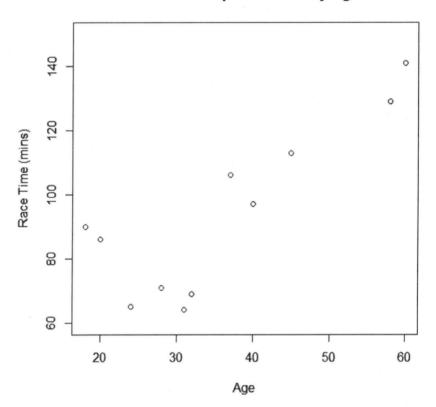

Figure 3-12. Scatter plot of time to complete a race by age, excluding the data entry error

Figure 3-13 shows a box plot of the time taken to read a 300 page book by different school year groups. This highlights a few outliers, which can occur quite often, that are shown as circles; however these are clearly statistical outliers only due to human variation and should therefore be left in and used in the analysis.

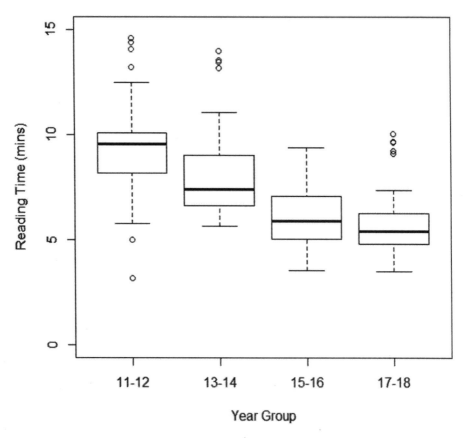

Figure 3-13. Box plot of reading time of a 300 page book by year group

It is important to remove data entry errors and technical errors so that the conclusions from the analysis have not been skewed by these points.

It is equally important to not remove genuine data, those points that are statistical outliers but are still within realistic bounds, as these need to be included to present the whole picture during the conclusions section.

Distribution

There are many different distributions, or families, that data sets can take with arguably the most well-known, for continuous data at least, being the normal distribution, also called the Gaussian distribution.

Figure 3-14 shows an example of a normal distribution that has the nice bell curve shape. You should be able to see how a histogram can aid in determining whether your data can be classed as approximately normal. The "tails" on this plot actually extend from minus infinity to plus infinity. This will not be the same for other distributions.

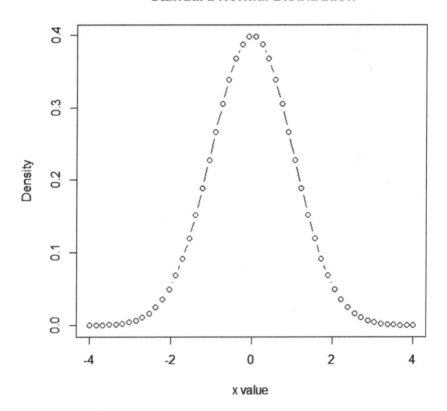

Figure 3-14. Standard normal distribution

When determining if your data follows a normal distribution the best method to use is drawing a graph rather than using formal normality tests, such as the Anderson–Darling or Shapiro–Wilk test. The most useful plot is a quantile-comparison plot, which also can be called a quantile-quantile plot or a Q–Q plot. This plot will draw the ordered values you have observed against theoretical expected values from a normal distribution.

Figure 3-15 shows an example of a Q–Q plot where we could assume normality in our data. Generally you want the points to line up nicely from the bottom left of the plot to the top right of the plot, along the $y = x$ line.

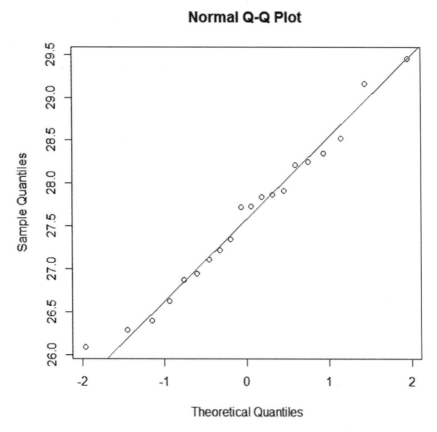

Figure 3-15. Quantile–Quantile (Q–Q) plot

You soon will notice an issue if there is curvature in the line or if the line of points is on a more horizontal line, in which case a transformation may help, see more in Chapter 4.

The reason formal tests are not recommended is that they can be misleading, especially with a large sample size. For example, in Figure 3-16 using the Q–Q plot you would be able to assume normality on that data, however the formal tests suggest strong non-normality. This is purely down to the fact that as there is such a large sample size, those slight deviations at the tails are enough for the formal test to say "non-normality."

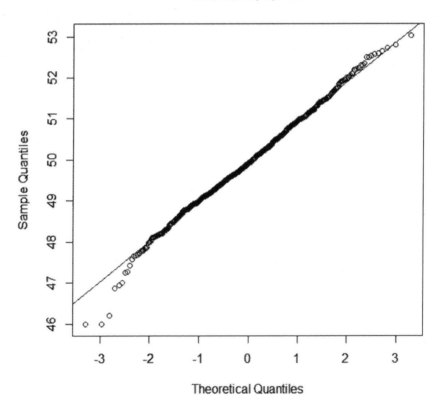

Figure 3-16. Q–Q plot of a large sample size

Other common continuous data distributions include the exponential and the gamma distributions. The exponential distribution is a special case of the gamma distribution with a fixed shape parameter of 1. A good way to think about an exponential distribution is modeling the time until an event occurs. The "tails" on this plot extend from zero to infinity.

Figure 3-17 shows an example exponential distribution with a rate of 1, the higher the rate the more the line will curve into an "L" shape.

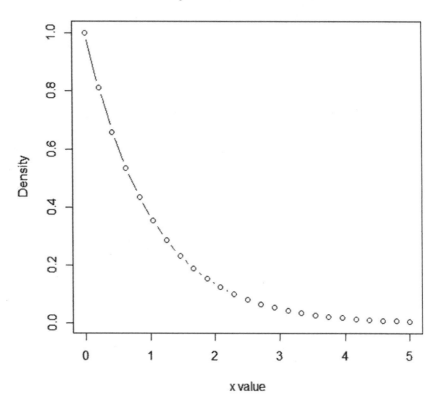

Figure 3-17. Exponential distribution with a rate of 1

The gamma distribution is a generalization of the exponential distribution and has a distinct shape, it is frequently used to model general waiting times, so modeling the time until the next *n* events.

Figure 3-18 shows an example gamma distribution with a rate of 1 and a shape of 2. Generally speaking, though it's not quite this simple, the higher the rate the more "squashed" the curve will become, and the higher the shape the further to the right the curve will move. The "tails" again extend from zero to infinity.

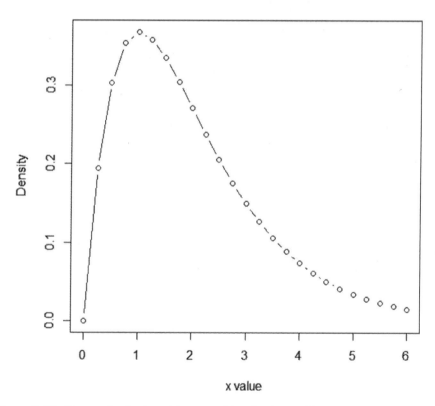

Figure 3-18. Gamma distribution with a rate of 1 and a shape of 2

The most commonly known discrete data distribution is the binomial distribution. This distribution is the probability of "success", which will be between 0 and 1 plotted against the number of trials.

Figure 3-19 shows an example binomial distribution with a size of 50 and a probability of 0.1, the size is just the number of trials, and the probability is the probability of success. The "tails" for a binomial distribution extend from zero to *n*, which in this case is 50.

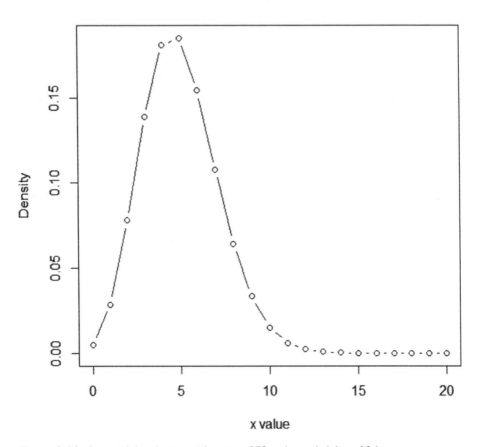

Figure 3-19. Binomial distribution with a size of 50 and a probability of 0.1

Another common distribution for discrete data is the Poisson distribution. This distribution is used for count data plotted against time, generally with there being a much larger count at the lower end of the plot.

Figure 3-20 shows an example Poisson distribution with a lambda of 3, the lambda dictates where the central peak of the curve will be. The "tails" extend from zero to infinity.

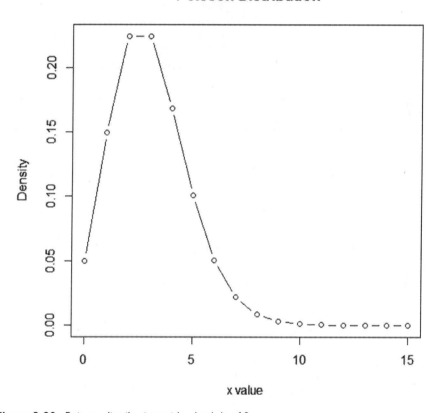

Figure 3-20. Poisson distribution with a lambda of 3

The list is by no means exhaustive, but those are the most commonly used distributions. Ideally if the data can be assumed to be normally distributed then that is the easiest way forward, however that isn't always the case and as such the relevant distribution needs to be identified.

Tests

Once you have progressed through each step of the EDA process and have learned more about the data, then you can move to deciding which tests, and descriptive statistics, would be appropriate. The key is not to jump straight to this step as valuable time and money may be wasted if the analysis is completed including a data entry error, or if the wrong analysis is performed due to not thinking about which data types are being used.

Continuous Data

If your continuous data approximately follows a normal distribution the types of tests used are called parametric tests. If the distribution is not normal and a transformation either doesn't help or is not appropriate then nonparametric tests may be more appropriate.

Generally speaking, nonparametric tests have less power than parametric tests, that is, there's a higher risk of missing a real effect by using nonparametric tests. This is due to the fact that nonparametric tests are distribution free and have to be conservative to account for this fact.

There are nonparametric equivalents for each parametric test and these will be shown alongside each other in the later chapters. The key difference between the tests, in addition to the power mentioned above, is that parametric tests use the means and nonparametric tests use the medians of the data.

In terms of satisfying the normality assumptions, data can actually be non-normal while still being applicable to parametric tests. There are some general rules of thumb to note: each case should be visually investigated to confirm whether appropriate to use parametric tests. As long as the sample size is above 15 for each group and the data is only slightly skewed, parametric tests can be used; or if the sample size is very large and the data clearly doesn't follow another distribution, parametric tests can be used.

Nonparametric tests are commonly used when there is a very small sample size due to the fact that the distribution won't be able to be identified. They also can handle statistical outliers and ranked data quite well. The downside of some nonparametric tests is that they assume that the groups have equal variation, which may not always be an appropriate assumption.

Discrete Data

To add slightly more confusion to the terms, if you had binary data that is clearly not continuous, then the type of tests you would use are parametric tests based on the binomial distribution. There are also parametric tests for Poisson and other distributions.

Basically parametric tests refer to traditional tests relevant to the data type, and nonparametric tests refer to tests used when the data violates the assumptions to be able to use the traditional tests. Most of the time parametric and nonparametric are associated with continuous data (and normality) only, but it's worth noting that these terms apply to all data types.

Thought needs to be given as to which type of tests are most appropriate to the data set you have collected, thinking about things such as data types, sample size, skewness, and variation. You also need to remember that all tests have assumptions, even nonparametric tests, and they need to be satisfied.

Summary

This chapter investigated the sections that make up exploratory data analysis (EDA), which should be performed before undertaking any type of statistical analysis. It was split into five sections corresponding to the five steps shown in the EDA process in Figure 3-1.

The first was identifying the data types for each variable in the data set as this will lead to the second section of plotting the data.

Plotting the data is dependent on which data types you have, as this dictates which plots can be drawn. This section showed a selection of commonly used plots and described the information to be gained from each. One main point in this section was that the plots drawn in EDA are for familiarity with the data only; Chapter 9 focuses on creating clear plots to deliver the main messages to the customer.

The third section discussed outliers in the data and that there are two types of outliers; data entry errors, which should be removed and statistical outliers, which should remain in the data during analysis.

The next section related to the family of the data, or the distribution. It listed the most common distributions used for continuous data and discrete data with an emphasis on checking for normality. It also highlighted that when checking for normality, plots such as the Q–Q plot should be used instead of formal tests.

The final section discussed the use of parametric and nonparametric tests and the benefits and downfalls of each of these along with the fact that these terms do not just apply to continuous data.

Chapter 4 looks at the descriptive statistics that can be gleaned from the data, again before testing is undertaken. It defines the difference between samples and populations, explains the different measures of shape, location, and spread; shows how to transform non-normal continuous data; highlights descriptive statistics for binary data; and also looks at what correlation can and can't tell you.

Descriptive Statistics

What Can the Data Tell Me?

The aim of calculating descriptive statistics is to summarize the sample you have collected. The key thing here is that these values or plots correspond only to the sample, there is no uncertainty and therefore the results can't be generalized to describe the population that step is the basis of Chapter 5.

Figure 4-1 shows the relationship between samples and populations with the following description:

- Top left: the population consists of the entirety of every-thing you are interested in, for example, all of the lupine flowers in the world.

- Top right: if the height of all those lupines were mea-sured, a probability distribution of the results could be drawn. These results wouldn't need any uncertainty, so the descriptive statistics reported would be the correct answer, until more grew or died.

- Bottom left: you clearly cannot measure the population of lupines, so you take a representative sample, defined by power and sample size calculations.

© Victoria Cox 2017
V. Cox, *Translating Statistics to Make Decisions*, DOI 10.1007/978-1-4842-2256-0_4

- Bottom right: this sample will produce a probability distribution and therefore descriptive statistics that you hope will represent the population. As mentioned previously though, the descriptive statistics can only be used to describe the sample and not infer the population, not until uncertainty has been introduced.

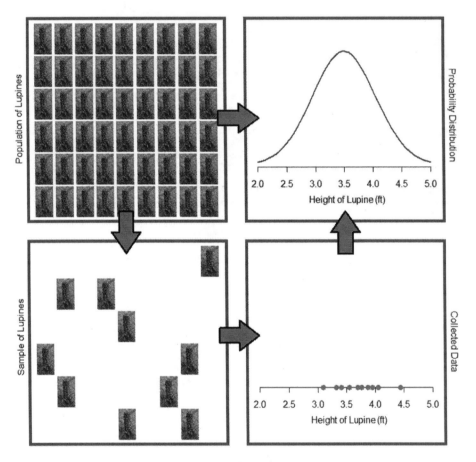

Figure 4-1. Relationship between samples and populations

Exploratory data analysis (EDA), which is mainly drawing plots, can sometimes be included in descriptive statistics due to the fact that it involves summarizing the collected data only. For the rest of the chapter the term *descriptive statistics* will refer to the numerical values and not EDA.

The majority of descriptive statistics are used to describe continuous data sets. Some of the methods can be used for discrete or qualitative data but not all; although in both cases they can be misused. The following sections explore the available options for the different data types.

Continuous Data

The three main characteristics that are of interest when dealing with continuous data are the shape, the location, and the spread of the data. The shape investigates the distribution of the data, the location looks at the center point, and the spread examines the dispersion of the data.

Shape

The shape of the data dictates the distribution of the sample, with the ideal shape being the bell curve corresponding to normality. However the shape also can inform you about symmetry through the skewness and kurtosis.

Skewness

Skewness measures the symmetry of the sample data. The recommended way of discerning symmetry is by plotting the data; if the tail is longer to the left and the hump is to the right then this is termed a *negative skew* or left skewed, and if the tail is longer to the right and the hump is to the left then this is termed a *positive skew* or right skewed, which is highlighted in Figure 4-2.

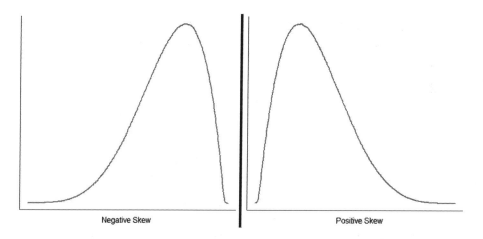

Negative Skew   Positive Skew

Figure 4-2. Visual examples of skewness

If the sample is negatively skewed, the median will be larger than the mean, and if it's positively skewed, the mean will be larger than the median (for definitions, see the Location section).

Skewness also can be defined in numerical terms; for the two graphed examples in Figure 4-2 the skewness would be –0.6 and +0.6, respectively, as they are mirror images of each other.

A perfectly symmetrical distribution would have a skewness of 0.

A general rule of thumb is,

- If the sample skewness is smaller than −1 then it is highly negatively skewed.

- If the sample skewness is larger than −1 but smaller than −0.5 then it is negatively skewed.

- If the sample skewness is larger than 0.5 but smaller than 1 then it is positively skewed.

- If the sample skewness is larger than 1 then it is highly positively skewed.

However you can see how it is more useful to show a plot of the skewness as opposed to quoting the value of skewness.

Kurtosis

Kurtosis is used less often than skewness, one of the reasons being it is harder to interpret. The best way of describing kurtosis is to think about the peak of the sample data, with a sharper, higher peak meaning a higher level of kurtosis and a flatter, lower peak meaning a lower level of kurtosis. However in terms of calculations, *kurtosis* is defined using the tails of the data more than the "peak" of the data. Due to this, it will be more affected by outliers than skewness would be. Figure 4-3 shows an example of positive kurtosis and negative kurtosis.

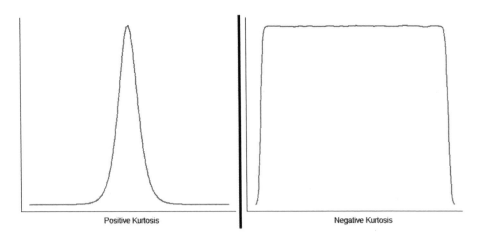

Positive Kurtosis

Negative Kurtosis

Figure 4-3. Visual examples of kurtosis

Kurtosis also can be defined in numerical terms; however excess kurtosis, which is (kurtosis − 3), is quoted more often, the reasoning behind this is given later. For the two graphed examples in Figure 4-3, the excess kurtosis would be 1.2 (kurtosis of 4.2) and −1.2 (kurtosis of −4.2), respectively.

The kurtosis for a normal distribution is 3, and the excess kurtosis is therefore 0; this is generally why excess kurtosis is quoted as 0 and is a more useful reference point. An excess kurtosis of 0 is also referred to as *mesokurtic*.

A general rule of thumb is as follows:

- If the sample excess kurtosis is smaller than 0 then it has negative kurtosis or is *platykurtic*.

- If the sample excess kurtosis is larger than 0 then it has positive kurtosis or is *leptokurtic*.

However again you can see how it is more useful to show a plot of the kurtosis as opposed to quoting the value of kurtosis or excess kurtosis.

Transformations

If the data is skewed often the first thing that should be tried is a transformation of the data as this could mean being able to use parametric tests on the sample.

For example, consider the data in Figure 4-4, which are weekly temperatures (°C) in the United Kingdom. The two plots shown are a histogram and a quantile-comparison plot.

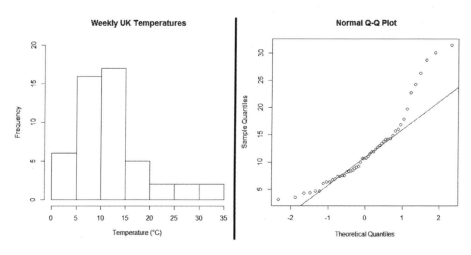

Figure 4-4. Skewed data plots

This data cannot be assumed to follow a normal distribution, and as such parametric tests could not be used. One of the most common transformations for positively skewed data is the $\log_{10}$ transformation. This is because it works on a lot of similarly shaped data sources and it's easy to explain: $\log_{10}(10)$ is 1, $\log_{10}(100)$ is 2, $\log_{10}(1,000)$ is 3, and so forth. Essentially it squashes the extreme values in the right tails closer to the left making the data much more symmetrical. Figure 4-5 shows the histogram and quantile-comparison plot for the $\log_{10}$ transformed data.

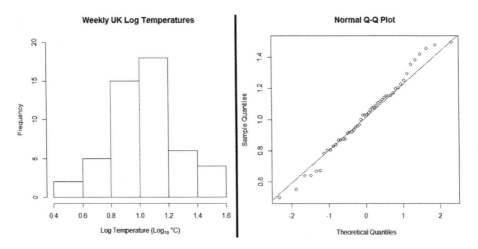

Figure 4-5. Transformed data plots

This data is clearly more symmetrical and can be assumed to follow a normal distribution, so parametric tests could be used.

When quoting descriptive statistics for transformed data remember that these have been calculated using the transformed scale, so it either needs to be made clear that log, or equivalent, values are being quoted or, preferably, the descriptive statistics should be back transformed to the original scale.

Figure 4-6 shows the R code for common transformations, some notes of their use, and back transformations with examples.

Used for	Method	Transformation	Back Transformation	Notes	Example
Positively skewed data	Log	log10(x)	10^y	No zeros No negative values	log10(5) = 0.699 10^0.699 = 5
	Natural log	log(x)	exp(y)	No zeros No negative values	log(7) = 1.946 exp(1.946) = 7
	Cube root	x^(1/3)	y^3	Weaker than log transformations	-8^(1/3) = -2 (-2)^3 = 8
	Square root	sqrt(x)	y^2	No negative values Weaker than log and cube root transformations	sqrt(9) = 3 3^2 = 9
	Reciprocal	1/x	1/y	No zeros No negative values	1/4 = 0.25 1/0.25 = 4
Negatively skewed data	Use the "back transformations" above for the transformations, then use the "transformations" above to do the back transformations				As above
Other	Arcsine	asin(sqrt(x))	sin(y)^2	Used with proportions For percentages use asin(sqrt(x/100)) then (sin(y)^2)*100	asin(sqrt(0.6)) = 0.886 sin(0.886)^2 = 0.6
	Logit	log(x/(1-x))	exp(x)/(1 + exp(x))	Used with proportions	log(0.4/(1-0.4)) = -0.405 exp(-0.405)/(1+exp(-0.405)) = 0.4

Figure 4-6. Common transformations

Data transformations, if required, should be carried out on the data before conducting any testing as they are designed to satisfy symmetry/normality assumptions and not to "force a significant result."

Location

The location is concerned with where the peak of the data lies. There are three common measures of location: the mode, median, and mean. For data that follows a normal distribution the median and the mean will be roughly the same.

There is also the weighted mean that can be a very useful measure of location for either unequal sample sizes or to amend population representation within the sample.

Mode

The mode is the most common value in the data set, for example the mode in Figure 4-7 would be 2.

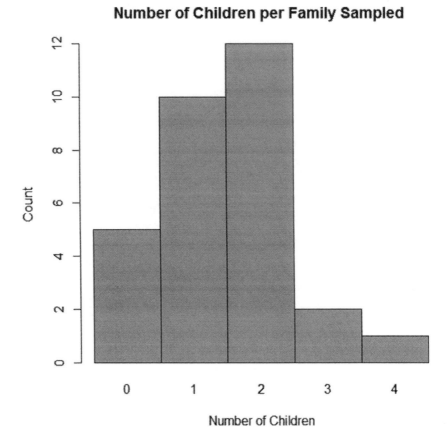

Figure 4-7. Histogram of the number of children per family sampled

Strictly speaking the mode is used for discrete data; however it is always grouped with the mean and the median when describing measures of location. The mode can be calculated for continuous data however it is not recommended due to the fact there is unlikely to be any repeated values.

Median

The median is the center value of the data, the values are ordered and the middle value is the median. If there are an odd number of values the median is just the center value, if there are an even number of values the median is the average of the middle two values.

Example 4.1 shows how to calculate the median for both skewed data and normally distributed data.

EXAMPLE 4.1

Calculate the median from skewed data (data3), then calculate the median for simulated normal data (data4).

```
# Create some skewed data
data3 = c(6.61, 7.88, 7.54, 8.08, 8.07, 7.2, 6.81, 6.45, 7.34, 6.27,
          6.19, 6.63, 19.98, 7.36, 7.18, 7.86, 7.33, 19.02, 8.03,
          8.04, 7.16, 7.14, 7.61, 7.3, 6.75, 6.71, 20.23, 7.67, 6.89,
          7.15, 7.52, 8.17, 7.55, 6.8, 19.72, 6.43, 8.05, 6.88, 13.08,
          10.16)
```

```
# Calculate the median for the skewed data
median(data3)
```

[1] 7.35

```
# Simulate some normally distributed data
data4 = rnorm(40, mean = 7, sd = 0.75)
```

```
# Calculate the median for the normally distributed data
median(data4)
```

[1] 6.951663

Note that answers may vary for the normal data median due to the fact that the **rnorm()** command generates random data from the normal distribution with a mean of 7 and a standard deviation of 0.75.

The median will always be the center of the data to emphasize this we will use the same data examples in the next section.

Mean

The mean, also called arithmetic mean, is the average of the results; the values are all added together and then divided by the number of samples. When the data follows a normal distribution the mean will be approximately equal to the median and therefore the center of the data.

It is important to note that when the data is skewed the mean will not represent the center of the data as it will be heavily influenced by the extreme values. In these cases it would be better to quote the median.

As a side note, there is also a geometric mean that can be used on lognormal data or data from different ranges, such as 2/5 and 10/100. It is calculated by the nth root of the product of the numbers, so for example to calculate the geometric mean for three numbers you would multiply the three together and then take the cube root of that answer.

Example 4.2 shows how to calculate the mean for both previous sets of data and then compares these results to the medians.

EXAMPLE 4.2

Calculate the mean for the skewed data (data3), then calculate the mean for the simulated normal data (data4).

 # Calculate the mean for the skewed data
mean(data3)

[1] 8.721

 # Calculate the mean for the normally distributed data
mean(data4)

[1] 6.989539

If you compare the mean results with the previous median results you can see that the skewed data gave different results: 7.35 compared to 8.72, whereas the normal data gave similar results: 6.95 compared to 6.99.

Figure 4-8 shows the plotted data for both the skewed data and the normally distributed data with the medians and means highlighted.

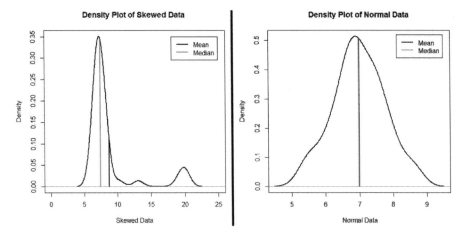

Figure 4-8. Density plots of skewed and normal data

There may be cases where you need to calculate a mean for each group as opposed to an overall mean, see Example 4.3.

EXAMPLE 4.3

Calculate the mean temperature (°F) for each location and an overall mean for the data given below.

```
# Create the data
Temp = c(72, 70, 71, 70, 90, 88, 87, 83, 75, 89, 91, 79, 93, 74, 86,
         84, 86, 90, 92, 75, 74, 87, 83, 81, 90, 50, 61, 59, 51, 55,
         58, 52, 52, 56, 55, 52, 61, 54, 56, 59, 57, 53, 72, 67, 83,
         76, 80, 65, 85, 77, 83, 71, 84, 78, 74, 65, 72, 75, 79, 76,
         69, 78, 71, 74, 65, 69, 66, 76, 70, 79, 66, 69)
Groups = c(rep("A", 25), rep("B", 17), rep("C", 30))
data5 = data.frame(Temp, Groups)

# Calculate means for each group
tapply(data5$Temp, data5$Groups, mean)
```

```
      A          B          C
82.40000    55.35294    73.80000
```

```
# Calculate the overall mean
mean(data5$Temp)
```

```
[1] 72.43056
```

Here you can see that the mean temperatures for each location are very different to each other and it would be incorrect to have just calculated an overall mean.

Careful consideration needs to be given to using the mean, it may not always be the most appropriate statistic to quote; if the data is skewed then a median would be better for summarizing the location of the center of the data.

Weighted Mean

In some cases a weighted mean may be preferable as it accounts for unequal sample sizes. In Example 4.3 there was no need to combine the temperature data as the locations were independent, however imagine if the same data was test scores from three classes and we only had the means.

Say you want to calculate an overall average score from a set of group means, but there are unequal sample sizes that need to be accounted for. To calculate a weighted mean you need to find the proportion the sample size of each group represents in relation to the total number. These values are then multiplied by the group means and added together to give an overall weighted mean, see Example 4.4.

EXAMPLE 4.4

Calculate the overall mean test score from the group means, then calculate the weighted mean test score for the data and compare to the arithmetic mean as if all data was available.

```
# Change the name of the data
data5$Scores = data5$Temp
```

```
# Calculate the means per group
tapply(data5$Scores, data5$Groups, mean)
```

```
        A          B          C
 82.40000   55.35294   73.80000
```

```
# Calculate the mean ignoring different group sizes
mean(c(82.4, 55.35294, 73.8))
```

```
[1] 70.51765
```

```
# Recall the sample size of each group
tapply(data5$Scores, data5$Groups, length)
```

```
 A  B  C
25 17 30
```

```
# Calculate the proportions of each group
25/(25+17+30); 17/(25+17+30); 30/(25+17+30)
```

```
[1] 0.3472222
[1] 0.2361111
[1] 0.4166667
```

```
# Calculate the weighted mean
(0.3472222*82.4) + (0.2361111*55.35294) + (0.4166667*73.8)
```

```
[1] 72.43056
```

```
# Calculate the arithmetic mean from the raw data
mean(data5$Scores)
```

```
[1] 72.43056
```

Here the overall mean of the means would have been 70.52, which would be incorrect as it doesn't account for the different group sample sizes. Accounting for the unequal class sizes brings the weighted mean up to 72.43, which is equal to that of the arithmetic mean if all data was available.

Using a weighted mean gives more influence to the larger groups, therefore thought needs to be given to whether this is appropriate for the data you are dealing with.

Spread

The spread investigates the dispersion of the data, essentially looking at how variable the sample data is. The most common measurements of spread are standard deviation, variance, range, quantiles and percentiles, and interquartile range (IQR).

Other useful measurements for dispersion include the median absolute deviation (MAD) and the coefficient of variation (CV).

Standard Deviation and Variance

The standard deviation is the most common statistic quoted for the dispersion of data. However it is only useful when the data follows a normal distribution. If the data is skewed or has extreme values, the standard deviation will be misleading.

The standard deviation is measured using the same units as the data, which makes it slightly easier to interpret. The larger the standard deviation is the larger the dispersion of the data. Figure 4-9 highlights this using data showing time to complete a task.

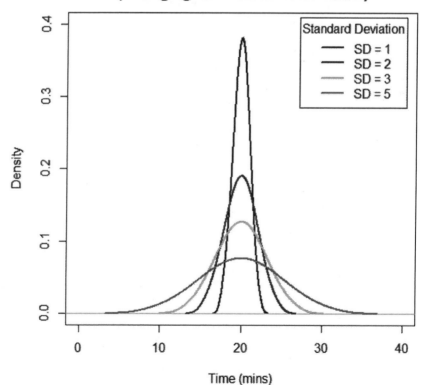

Figure 4-9. Varying standard deviations

In addition if the data follows the normal distribution well, it can be said that roughly 68% of the sample values are contained within 1 standard deviation from the mean, 95% within 2 standard deviations, and 99.7% within 3 standard deviations, see Figure 4-10.

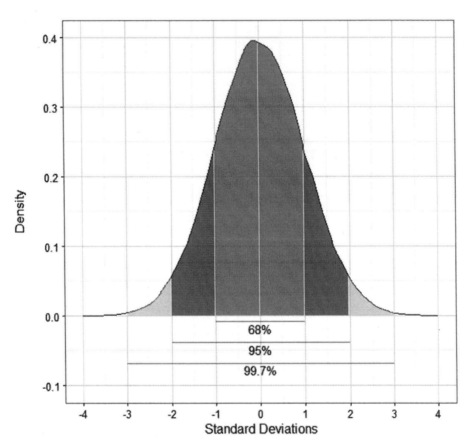

Figure 4-10. Data contained in varying standard deviations from the mean

However if the data is even slightly skewed or has extreme values, these calculations will be wrong. It's also worth noting that quoting these values have no associated statistical confidence and are not measurements of uncertainty, which will be discussed in Chapter 5.

Variance is simply the standard deviation squared. The issues with using the variance as a statistic for dispersion instead of the standard deviation are mainly twofold. First is that the result is not in the same units as the original data and therefore it is hard to reference back to the values. Second is that as the values are squared, it will give more emphasis to the extreme data points, which may not be representative.

To calculate the standard deviation and the variance in R, the commands are *sd(x)* and *var(x)*, respectively, where x represents a column of continuous data.

Range

The range looks at the complete spread of the data and is calculated as the maximum value minus the minimum value. It isn't always the most helpful statistic for dispersion. However it can inform you if there are extreme values, especially if your subject area has thresholds that cannot be crossed and if there have been data entry errors as well.

The range can be quoted as a single value or as the range itself, see Example 4.5.

EXAMPLE 4.5

Calculate the range for both the earlier skewed data and the simulated normal data.

```
# Calculate the range for the skewed data
range(data3)
```

[1] 6.19   20.23

```
# Calculate the range as a single value for the skewed data
max(data3) - min(data3)
```

[1] 14.04

```
# Calculate the range for the normally distributed data
range(data4)
```

[1] 5.342806   8.605029

```
# Calculate the range as a single value for the normally distributed data
max(data4) - min(data4)
```

[1] 3.262223

Here you can see that the range is much larger for the skewed data, and this is linked to the extreme values in the data set.

The range is very difficult to estimate as it essentially contains 100% of the data and a sample will rarely include the extremes.

Quantiles and Percentiles

Quantiles divide the data set up into equal groups, the number of which can be any positive value.

Many quantiles have a specific name with regard to the number of sections the data is divided into; Figure 4-11 shows these names.

Quantiles	Name
2	Median
3	Tertiles
4	Quartiles
5	Quintiles
6	Sextiles
7	Septiles
8	Octiles
10	Deciles
12	Duo-deciles
16	Hexadeciles
20	Ventiles
100	Percentiles
1000	Permilles

Figure 4-11. Special quantile names

Quantiles can split the data into as many intervals as required. For example, if the data is split into permilles, then there would be quantile values for every 0.1% of the data. As an aside, percentiles also can be known as centiles.

When calculating quantiles in R, the length specified needs to be the number you need plus one. For example if you need quintiles, which is five sections, you would need to specify the length as six. This is due to the fact that it will give you the values at 0%, 20%, 40%, 60%, 80%, and 100%; so while there it has divided the data into five sections, it will quote the six values.

Quartiles are the most well-known and most often used quantiles, they split the data into four intervals with generally only three being quoted:

- Q1: the first quartile is the value below which is 25% of the data.

- Q2: the second quartile is the value below which is 50% of the data, it is also the median.

- Q3: the third quartile is the value below which is 75% of the data.

Percentiles split the data into the percentages of the data, for example the 20th percentile is where 20% of the data would fall below the given value. Percentiles are usually only whole percentages, however there is clearly over-lap as the 20th percentile is the equivalent to the 1st quintile.

Example 4.6 shows how to calculate quantiles, quartiles, and percentiles using the previous data sets.

EXAMPLE 4.6

Calculate the quintiles, the quartiles, and the 95th and 99th percentiles for the skewed and simulated normal data.

```
# Calculate quintiles for both the skewed and the normally
distributed data
quantile(data3, prob = seq(0, 1, length = 6))
quantile(data4, prob = seq(0, 1, length = 6))
```

0%	20%	40%	60%	80%	100%
6.190	6.790	7.192	7.574	8.072	20.230

0%	20%	40%	60%	80%	100%
5.342806	6.409402	6.840612	7.145605	7.575320	8.605029

```
# Calculate quartiles for both the skewed and the normally
distributed data
quantile(data3); quantile(data4)
```

0%	25%	50%	75%	100%
6.1900	6.8625	7.3500	8.0425	20.2300

0%	25%	50%	75%	100%
5.342806	6.543953	6.951663	7.472544	8.605029

```
# Calculate 95th and 99th percentiles for the skewed and the normally
# distributed data
quantile(data3, prob = c(0.95, 0.99))
quantile(data4, prob = c(0.95, 0.99))
```

95%	99%
19.7330	20.1325

95%	99%
8.150095	8.601126

You can see how much closer and evenly spread the normal data results are compared to the skewed data results. In addition you can see that the skewed data is fairly consistent in dispersion for up to 80% of the data that means that the extreme values are all at the higher end of the data, which is confirmed by previously plotting the data.

Technically there is always a Q0 that would be the equivalent of the minimum. However it is rarely reported as such, similar for the last quantile, which is the maximum (which for quartiles would be Q4). However, these are generally left as the minimum and maximum values or the range.

IQR and SIQR

The interquartile range (IQR) represents the middle 50% of the data, that is, Q3 minus Q1. It is less affected by outliers and skewed data compared to the standard deviation or variance.

The IQR can be quoted as a single value or as the range Q1 to Q3 the same as when quoting the range. The IQR is the main section of a box plot, with the top line being Q3 and the bottom line being Q1. Figure 4-12 shows a box plot with the sections labeled.

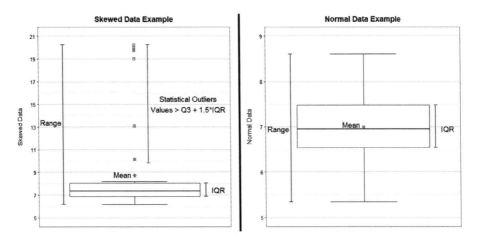

Figure 4-12. Box plots of the skewed and normal data

As described in Chapter 3, the whiskers will only represent the range if there are no statistical outliers. Here it also highlights that the mean and median are very close together for the normal data, but are quite far apart for the skewed data, in fact the mean is above Q3 for the skewed data.

A less commonly used measure of spread is the semi-interquartile range (SIQR), which represents half the IQR. It is more robust against skewed data than the standard deviation but is not necessary if the data follows a normal distribution.

To calculate the IQR and the SIQR in R, the commands are **IQR(x)** and **IQR(x)/2**, respectively, where x represents a column of continuous data.

MAD

The median absolute deviation (MAD) is another good measure of dispersion in the data, and it is more robust to outliers than the standard deviation.

The MAD is calculated by finding the positive difference between each value and the median of the data, and then finding the median of all those results.

Example 4.7 walks through the method for calculating the MAD and then shows a shortcut using an R command.

EXAMPLE 4.7

Calculate the MAD for the following data: 1, 4, 3, 5, 6, 2, 4, 2, 3, 4.

Rearrange the order of the data:
1, 2, 2, 3, **3, 4,** 4, 4, 5, 6

Calculate the median of the values:
3.5

Subtract the median from all values in the data:
(1 - 3.5), (4 - 3.5), (3 - 3.5), (5 - 3.5), (6 - 3.5), (2 - 3.5), (4 - 3.5), (2 - 3.5), (3 - 3.5), (4 - 3.5)

This will equal:
-2.5, 0.5, -0.5, 1.5, 2.5, -1.5, 0.5, -1.5, -0.5, 0.5

Take absolute values:
2.5, 0.5, 0.5, 1.5, 2.5, 1.5, 0.5, 1.5, 0.5, 0.5

Rearrange the order of the data:
0.5, 0.5, 0.5, 0.5, **0.5, 1.5,** 1.5, 1.5, 2.5, 2.5

Calculate the median of the values:
1

This can be done simply in R with the following command:

```
# Create the data
x = c(1, 4, 3, 5, 6, 2, 4, 2, 3, 4)

# Calculate the MAD
mad(x, constant = 1)
```

[1] 1

The MAD for the data is 1, as this is small it suggests that the median value is a good representation for the other values in the data.

The MAD can also be used to highlight outliers by calculating the absolute deviation from the median for each point, then dividing them by the MAD. This will show the distance of the values from the center of the data in terms of the MAD, see Example 4.8.

EXAMPLE 4.8

Determine if there are any outliers from the previous data set. Repeat the same question replacing the value of 1 with 10.

```
# Create the data
x = c(1, 4, 3, 5, 6, 2, 4, 2, 3, 4)

# Calculate the distance from the center in terms of the MAD
abs(x - median(x)) / mad(x, constant = 1)
```

[1] 2.5 0.5 0.5 1.5 2.5 1.5 0.5 1.5 0.5 0.5

```
# Create the data swapping 1 for 10
x2 = c(10, 4, 3, 5, 6, 2, 4, 2, 3, 4)

# Calculate the distance from the center in terms of the MAD for the new data
abs(x2 - median(x2)) / mad(x2, constant = 1)
```

[1] 6 0 1 1 2 2 0 2 1 0

For the first set of data the values are all quite similar and not that different to the MAD of 1, so we can conclude there are no statistical outliers. For the second set of data, which still has a MAD of 1, the new value of 10 is clearly a statistical outlier as the distance of 6 is much larger than all the other values and the MAD itself.

There can be some confusion as MAD also can be used to refer to the mean absolute deviation, which I call the average absolute deviation (AAD) to avoid this issue.

The AAD will tell us on average how far the values are from the mean, and the MAD will tell us the median distance of the values from the median. If the data is skewed the MAD is preferable to the AAD, however the results for both will be similar if the data follows a normal distribution as Example 4.9 highlights.

EXAMPLE 4.9

Calculate the MAD and the AAD for the skewed data and the normal data and determine if there are any statistical outliers.

```
# Calculate the MAD for the skewed data
MADs = mad(data3, constant = 1); MADs
```

[1] 0.575

```
# Calculate the AAD for the skewed data
AADs = mean(abs(data3 - mean(data3))); AADs
```

[1] 2.4932

```
# Calculate the MAD for the normally distributed data
MADn = mad(data4, constant = 1); MADn
```

[1] 0.5043415

```
# Calculate the AAD for the normally distributed data
AADn = mean(abs(data4 - mean(da ta4))); AADn
```

[1] 0.6003418

```
# Calculate the distance from the center in terms of the MAD for the
skewed data
abs(data3 - median(data3))/ MADs
```

```
[1]  1.28695652   0.92173913   0.33043478  1.26956522   1.25217391
[6]  0.26086957   0.93913043   1.56521739  0.01739130   1.87826087
[11] 2.01739130   1.25217391  21.96521739  0.01739130   0.29565217
[16] 0.88695652   0.03478261  20.29565217  1.18260870   1.20000000
[21] 0.33043478   0.36521739   0.45217391  0.08695652   1.04347826
[26] 1.11304348  22.40000000   0.55652174  0.80000000   0.34782609
[31] 0.29565217   1.42608696   0.34782609  0.95652174  21.51304348
[36] 1.60000000   1.21739130   0.81739130  9.96521739   4.88695652
```

```
# Calculate the distance from the center in terms of the AAD for the
skewed data
abs(data3 - mean(data3))/ AADs
```

```
[1]  0.8467030  0.3373175  0.4736884  0.2570993  0.2611102
[6]  0.6100594  0.7664848  0.9108776  0.5539066  0.9830740
[11] 1.0151612  0.8386812  4.5158832  0.5458848  0.6180812
[16] 0.3453393  0.5579175  4.1308359  0.2771539  0.2731429
[21] 0.6261030  0.6341248  0.4456121  0.5699503  0.7905503
[26] 0.8065939  4.6161559  0.4215466  0.7343976  0.6301139
[31] 0.4817103  0.2210011  0.4696775  0.7704957  4.4115996
[36] 0.9188994  0.2691320  0.7384085  1.7483555  0.5771699
```

```
# Calculate the distance from the center in terms of the MAD for the
normally distributed data
abs(data4 - median(data4))/ MADn
```

```
[1]  0.09080554  1.26333645  0.05428266  1.61245704  1.51548306
[6]  1.05829284  3.27826681  0.48997753  0.86828270  1.22988689
[11] 0.57145605  1.45430428  0.11157321  0.45287171  1.02429604
[16] 0.25993697  0.05428266  3.25842509  0.09118028  3.19001510
[21] 2.06479142  2.65414407  0.28295510  0.55616482  0.79607567
[26] 1.68002831  1.06347187  0.94021016  2.04742818  2.83061775
[31] 0.05428266  2.32980034  0.84537759  0.64911573  0.17834543
[36] 0.68850967  1.72453982  0.97570396  1.12204528  2.11762467
```

```
# Calculate the distance from the center in terms of the AAD for the
normally distributed data
abs(data4 - mean(data4))/ AADs
```

```
[1]  0.13937556  0.99822635  0.10869304  1.41770062  1.21005234
[6]  0.82597115  2.69095033  0.47471618  0.66634542  0.97012569
[11] 0.41698415  1.28483798  0.15682228  0.44354394  0.79741075
[16] 0.15527991  0.01748833  2.67428149  0.01350901  2.74299226
[21] 1.79770247  2.29281207  0.30079859  0.53031950  0.73186634
[26] 1.34828525  0.95650337  0.72677100  1.65693441  2.44106594
[31] 0.01748833  1.89415260  0.77328441  0.48222532  0.21291699
[36] 0.64150116  1.51186031  0.75658901  1.00571033  1.71590582
```

Comparing the MAD and the AAD for the skewed data you can see there is quite a difference in results compared to the MAD and the AAD for the normal data, which produce similar values. When looking for outliers: the MAD for the skewed data confirms four extreme outliers and suggests two less extreme outliers, whereas the AAD really only picks out the four extreme outliers, which are highlighted in bold in the R output. There are no outliers in the normal data, which is to be expected.

CV

The coefficient of variation (CV) is very useful if you want to compare data sets with different units. Another common area of use is to compare dispersion at differing concentration levels.

The CV is calculated as the standard deviation divided by the mean, which means that the value is normalized and can be compared across different groups as it is "unitless." Due to this, the CV gives a measure of relative dispersion whereas the standard deviation gives a measure of absolute dispersion.

It is also commonly converted to a percentage by multiplying the CV by 100, to give the percentage coefficient of variation (%CV).

Example 4.10 shows how to calculate the %CV from the previous datasets and compares this to the standard deviations.

EXAMPLE 4.10

Compare the mean, standard deviation, and %CV for both the skewed data and the normal data.

 # Calculate the mean for both the skewed and the normally distributed data
mean(data3); mean(data4)

[1] 8.721
[1] 6.989539

 # Calculate the standard deviation for both the skewed and the normally distributed data
sd(data3); sd(data4)

[1] 3.893894
[1] 0.771864

 # Calculate the percentage coefficient of variation for both the skewed and the normally distributed data
(sd(data3) / mean(data3))*100; (sd(data4) / mean(data4))*100

[1] 44.64963
[1] 11.04313

The means for the skewed data and the normal data are similar to each other, but we can clearly see from the standard deviation that the skewed data is more variable than the normal data. However we can't directly compare these values. Calculating the %CV lets us do that; the skewed data has a %CV of 44.6% and the normal data a %CV of 11.0%, these values are now directly comparable and we can say that the skewed data is four times more dispersed than the normal data.

Discrete Data

There are fewer descriptive statistics when dealing with discrete and qualitative data due to the nature of the data. In general the only things that can be done are to summarize the counts, turn the counts into proportions or percentages, or calculate the mode.

For example, if you were collecting data on the colors of passing cars you couldn't calculate a mean or a standard deviation, you could however calculate the mode. Figure 4-13 shows some example calculations you could make on the data; the mode for this data would be "black." However the mode isn't always the most useful piece of information on its own as it may not be a good representation of the data, in the example below under a quarter of the cars are "black."

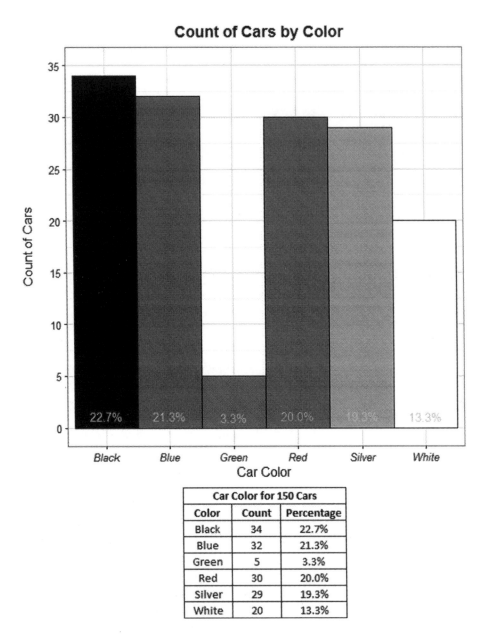

Figure 4-13. Displaying discrete data

It is worth noting that the more groups there are, the easier it is to see in a plot as opposed to tables of counts. Likert response data in particular is always clearer to view in a Likert plot as opposed to tables of counts. On the other hand if you just have a binary response for one group, this would be better presented in a table than a plot.

Bivariate Data

Bivariate data simply refers to data with two variables that are recorded. This can apply to both discrete and continuous data. The methods to calculate descriptive statistics that describe the relationship between the variables differ for the different data types. With discrete data a common approach is to create contingency tables and with continuous data a common approach is to look at correlation.

Contingency Tables

Contingency tables are a good way to display the counts for two discrete, or quantitative, variables. As with univariate data a good accompaniment to the tables is a bar chart.

Example 4.11 shows how to create a contingency table for discrete data.

EXAMPLE 4.11

Create a contingency table from the given data.

```
# Create the data
gender = rep(c("M","F"), each = 25)
smokes = sample(c(0,1), 50, replace = TRUE)
data6 = data.frame(gender, smokes)

# Create a contingency table
xtabs( ~ gender + smokes, data = data6)
```

```
        smokes
gender    0    1
     F   13   12
     M   15   10
```

Note that answers may vary as the **sample()** command randomly choses 0 or 1 values. This xtabs command can be run on variables with more than two levels as well as more than two variables.

Correlation and Covariance

The covariance between two continuous variables shows how the two variables are related in terms of a positive or a negative linear relationship.

This alone is not very useful, so the correlation coefficient is a normalized measurement to give more information. Correlation is calculated by dividing the covariance by the product of their standard deviations.

A general rule of thumb is as follows:

- If the correlation coefficient is close to −1 then the two variables are strongly negatively correlated.

- If the correlation coefficient is close to 0 then the two variables are either not correlated or very weakly correlated.

- If the correlation coefficient is close to 1 then the two variables are strongly positively correlated.

There are two important items to note when using and quoting the correlation coefficient:

- Correlation only implies a linear relationship between the variables. The variables may have a perfect relationship with each other; however if it's not on a straight line then the correlation won't recognize the relationship.

- Correlation does not mean causation—just because the data has a strong correlation does not mean that one variable has an effect on the other variable.

Figure 4-14 shows examples of differing correlations with the correlation coefficients.

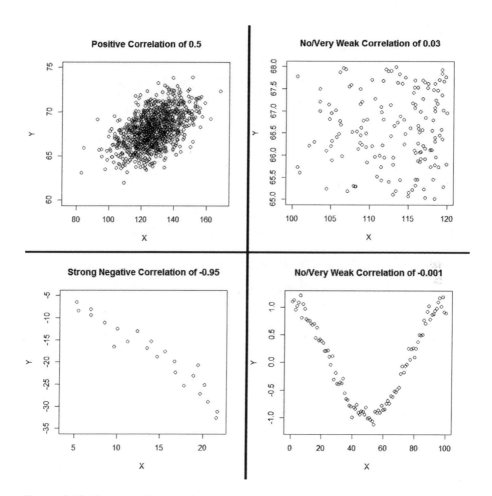

Figure 4-14. Examples of correlations

The top left box shows a positive correlation however it is very noisy data, which is why the coefficient is 0.5. The top right box shows no trends or patterns, which is why the coefficient is almost 0. The bottom left box shows a strong negative correlation with hardly any noise in the data, which is why there is a correlation coefficient of -0.95. The bottom right box clearly has a strong relationship between the variables, however the trend down on the left and up on the right cancel each other out which is why the coefficient is nearly 0, remember correlation only shows the linear relationship, which is not always useful to know.

To calculate the correlation in R use the command *cor(y = variable1, x = variable2)*, where variable1 and variable2 represent two continuous variables.

Summary

The beginning of the chapter described the relationship between samples and populations; then went on to investigate the different elements of descriptive statistics for continuous data, discrete data, and bivariate data.

The continuous data section was split into three main sections: shape, location, and spread.

The shape section looked at the statistics that could be used to describe the distribution and symmetry of the data, such as skewness and kurtosis. It also delved into transformations that can be used on asymmetrical data.

The location section identified the statistics that could be used to find the center of the data, such as the mode, median, and the mean, with explanations about when each is appropriate to use. It also showed how to calculate a weighted mean when sample sizes are unequal.

The spread section discussed the many statistics that could be used to explain the dispersion of the data, such as the standard deviation, variance, range, quantiles, percentiles, IQR, SIQR, MAD, AAD, and the CV. There are examples in each section, whether visual or exercises, showing the calculations for each along with their benefits.

Next was a small section on discrete data, as the descriptive statistics for this type of data are limited to counts, proportions, percentages, and the mode.

The final section related to bivariate data and was split into two main sections: contingency tables and correlation and covariance.

The contingency tables section showed how to create the tables, which are just an extension of summarizing counts in the discrete section.

The correlation and covariance section gave a brief description of covariance, and then went on to highlight how correlation should be interpreted.

Although this chapter was used to describe the sample data, Chapter 5 includes the uncertainty surrounding these estimates so statements can be made about the population that the sample was derived from; the main elements of the chapter will be confidence intervals, tolerance intervals, and prediction intervals.

Measuring Uncertainty

How Good Is the Data?

Any sample collected is done to obtain information representative of the population. And as such there will be uncertainty on the measurements taken due to not being able to test the entire population.

Uncertainty around any type of estimate should always be quoted to emphasize the level of statistical confidence in the results, to show whether the estimate can be relied on when making an evidence-based decision, and quite simply, to show the range of values the estimate of the population could fall within.

As mentioned in Chapter 4, quoting uncertainty as "95% of measurements fall within 2 standard deviations" is not always a correct statement, and it does not include statistical confidence. As such those measurements are not recommended, especially when it isn't difficult to produce uncertainty with statistical confidence.

The three most common types of uncertainty surrounding an estimate are confidence intervals, tolerance intervals, and prediction intervals. This chapter is split into three sections to describe each one.

© Victoria Cox 2017
V. Cox, *Translating Statistics to Make Decisions*, DOI 10.1007/978-1-4842-2256-0_5

Confidence Intervals

Confidence intervals are used when you are interested in the uncertainty around a sample estimate such as the mean.

Confidence intervals can be one-sided or two-sided, which means that you either quote one confidence limit from one end of the data or both limits from either end. The confidence interval values are called the lower confidence limit/bound and the upper confidence limit/bound. Therefore, for a one-sided confidence interval you would quote either the lower bound or the upper bound, whereas for a two-sided confidence interval you would quote both bounds

The decision about calculating a one-sided or two-sided interval should be decided during the experimental design stage, as your power or sample size calculation would have been carried out using this information. It would be incorrect to quote just one bound from a two-sided confidence interval calculation as the confidence level associated with the statement would be wrong.

The reason you may use a one-sided confidence interval is if you are only interested in one end of the data. For instance, you only care about how bad a detector is but not really how good it can be, in which case you would only want a lower bound.

Figure 5-1 shows the risk, which is the significance level, associated with both a one-sided and two-sided 90% confidence interval for continuous data.

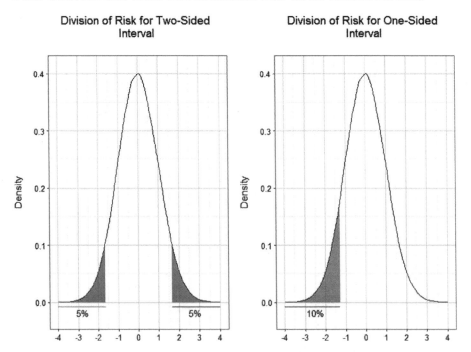

Figure 5-1. Division of risk for two-sided and one-sided confidence intervals

For a one-sided interval the risk is put at one end of the distribution with there being no information about the other end. Looking at Figure 5-1, 10% of the risk is at the lower end, so really the upper limit is infinity as there is no information at the upper end, if the risk was all at the higher end the lower limit would be minus infinity.

Thinking about multiple journey times to a location that involves heavily congested roads, you may not be interested in how quick the journey could be, but how long it may take. With a one-sided upper interval you would be able to say the average journey time will be no greater than z hours, but you can't make any kind of statement about how quick the journey could be.

For a two-sided interval the risk is split equally at each end, as you can see in Figure 5-1 there is 5% risk at each end. Therefore the lower bound here would be a smaller number than the lower bound of the one-sided interval.

With a two-sided interval you could state that the average journey time could be as long as x hours but also could be as quick as y hours. However the upper bound of x hours here would be a larger value than that of the z hours in the one-sided interval, as you are not as sure about each end in the two-sided case compared to the one-sided case.

The following sections look at calculating confidence intervals for continuous and binary data, with the information about sides also applying to binary data. The only difference is that the limits at the ends that are not of interest, in the one-sided interval case, are no longer minus infinity and infinity, but are 0 and 1 as binary data is bounded.

Continuous Data

To calculate a confidence interval for the mean of continuous data that approximately follows a normal distribution the following information is required:

- The mean of the sample.

- The variability of the sample: the standard deviation.

- The number of measurements in the sample: the sample size.

- The confidence level required or the level of risk that is acceptable.

- The interval of interest: one-sided or two-sided bounds.

Example 5.1 shows not only how to manually calculate a confidence interval in R to understand the steps involved, but also shows the shortcut using the **Rmisc** package. This shortcut only works for two-sided intervals unless you adjust the confidence level to compensate.

After creating the data, jump straight to "**# Load library**" if you are only interested in using the shortcut.

Part of the error calculation uses a critical value from the *t* distribution, which is based on the specified confidence level, the sample size, and whether the test is one-sided or two-sided.

EXAMPLE 5.1

Calculate a 90% two-sided confidence interval and a 90% one-sided lower confidence bound for the mean on the following data.

```
# Create some data
data7 = c(26.33, 27.31, 27.38, 26.63, 26.87, 26.67, 28.36, 28.52,
          26.91, 28.90, 27.99, 27.17, 28.32, 26.93, 26.93, 26.65,
          27.73, 26.93)

# Calculate the mean, standard deviation, and sample size for the data
# Assign the confidence level
x = mean(data7); s = sd(data7); n = length(data7); c = 0.90
x; s; n; c
```

```
[1] 27.36278
[1] 0.7585885
[1] 18
[1] 0.9
```

```
# Calculate the two-sided error
se = s/sqrt(n)
t2 = qt(c + (1 - c)/2, df = n - 1)
error2 = se*t2; error2
```

```
[1] 0.3110435
```

```
# Calculate the one-sided error
t1 = qt(c, df = n - 1)
error1 = se*t1; error1
```

```
[1] 0.2384096
```

```
# Calculate the lower and upper two-sided confidence intervals
# Calculate the one-sided lower bound
lower.2s.CI = x - error2; upper.2s.CI = x + error2
lower.1s.CI = x - error1
lower.2s.CI; upper.2s.CI
```

```
[1] 27.05173
[1] 27.67382
```

lower.1s.CI

```
[1] 27.12437
```

```
# Load library
library(Rmisc)
```

```
# Shortcut to calculate the two-sided interval
CI(data7, ci = 0.90)
```

```
   upper      mean      lower
27.67382   27.36278   27.05173
```

```
# Shortcut to calculate the one-sided interval with amendment
CI(data7, ci = 0.80)
```

```
   upper      mean      lower
27.60119   27.36278   27.12437
```

The two-sided confidence interval is (27.05, 27.67) or 27.36 ± 0.31 due to symmetry, and the one-sided lower bound is 27.12.

Note: If using the shortcut to calculate a one-sided interval you will need to double the significance level, here from 0.1 to 0.2, and make sure you only quote one side and not both.

In Example 5-1, if both the lower and upper one-sided bound had been calculated and incorrectly quoted as a two-sided interval, the values would be the same as calculating an 80% two-sided confidence interval, as 10% of risk would be at each end of the distribution totaling 20% risk overall.

Translating the output from a confidence interval is important as you need to remember the interval is around an estimate and not a proportion of the data. Using Example 5.1 you could state that you are "90% confident that the true mean lies within 27.05 and 27.67," or as the interval is symmetrical you could say that you are "90% confident that the true mean is 27.36 (± 0.31)," using an appropriate number of decimal points dependent on the required degree of accuracy.

The confidence level used will have a strong effect on the width of the confidence interval. For example, if you need a more precise interval around the mean then you could accept a lower confidence level. However this would result in a higher risk of being misled and that would result in a higher risk of the true mean being outside those interval values. If you need a high confidence level that would result in a lower risk of being misled, this would give a less precise interval.

The sample size used also will have a strong effect, as a larger sample size will provide a more precise interval whereas a smaller sample size will give a less precise interval.

In addition, the variation in the data also will have an effect with noisier data giving a larger and therefore less precise interval. Note that the variation of the data cannot be controlled.

Example 5.2 uses the same data as Example 5.1, data7, but this time looks at using a lower confidence level of 70%, then using a larger sample size of 25 to highlight the earlier points.

EXAMPLE 5.2

Calculate a 70% two-sided confidence interval for the original data, data7, and then a 90% two-sided interval for the following data, which has a larger sample size.

```
# Calculate the two-sided confidence interval at the 70% confidence level
CI(data7, ci = 0.70)
```

upper	mean	lower
27.55392	27.36278	27.17163

```
# Create the larger sized data
data8 = c(data7, 26.01, 28.33, 26.62, 26.99, 27.48, 27.74, 27.89)
```

```
# Calculate the two-sided confidence interval for the larger data
CI(data8, ci = 0.90)
```

upper	mean	lower
27.60178	27.34360	27.08542

The original confidence interval with a sample size of 18 and a confidence level of 90% was ±0.31, decreasing the confidence level to 70% reduces that to ±0.19; increasing the sample size to 25 with 90% confidence level (while maintaining a similar mean and variation) reduces that to ±0.26.

The calculations in Example 5-2 can be used to determine confidence intervals if one of the three assumptions in the following is satisfied,

- The data approximately follows a normal distribution.

- The data has a reasonable sample size and a symmetrical distribution.

- The data has a large sample size.

If none of these three points can be satisfied, one option is to use nonparametric confidence intervals however they can be quite complicated to compute. Although they do have fewer assumptions, they will give a less precise answer.

A better option to try first would be to transform the data using one of the transformations listed in Chapter 4. The key to remember when doing this is to back transform the information at the end to give the correct confidence interval. It is also worth noting that the interval will no longer be symmetrical around the arithmetic mean but it will be symmetrical around the geometric mean, which is also mentioned in Chapter 4.

Example 5.3 looks at an example where the data needs to be transformed for a confidence interval to be calculated.

EXAMPLE 5.3

Calculate a 95% two-sided confidence interval for the following skewed data and compare that to a confidence interval ignoring the normality assumption requirement.

```
# Create skewed data
data9 = c(9.2, 7.4, 10.7, 3.6, 4.3, 3.2, 14.2, 30.1, 15.7, 6.8, 8.9,
          9.1, 8.2, 7.5, 7.4, 14.9, 19.7, 26.3, 6.4, 14.2, 8.3, 6.9,
          8.5, 11.5, 22.7, 16.9, 31.4, 10.7, 17.9, 10.0)

# Transform the skewed data
data10 = log10(data9)

# Check normality - output plot ommited
qqnorm(data10); qqline(data10)

# Calculate the two-sided confidence interval for the transformed data
ci = CI(data10, ci = 0.95); ci
```

```
    upper        mean       lower
1.1183646   1.0246775   0.9309905
```

```
# Back-transform values to original scale
10^ci
```

```
    upper        mean       lower
13.133019   10.584675   8.530814
```

```
# Calculate the incorrect two-sided confidence interval for the
# skewed data
CI(data9, ci = 0.95)
```

```
    upper        mean       lower
15.189851   12.420000   9.650149
```

The back-transformed confidence interval is (8.53, 13.13), which would be a lower bound of 10.58 (- 2.05) and an upper bound of 10.58 (+2.55), which is clearly not symmetrical due to the skewed distribution.

If you compare that to the incorrectly calculated confidence interval of (9.65, 15.19), which is 12.42 (±2.77) you can see how different the two answers are and how the incorrect confidence interval could be misleading.

If you need to calculate confidence intervals in Excel use the following functions for two-sided and one-sided confidence intervals, respectively:

$$= AVERAGE(data) - T.INV.2T(alpha, (COUNT(data) - 1)) * STDEV.S(data) / SQRT(COUNT(data)).$$

Where data is the column of data you are using and where alpha is the significance level you are interested in, remember significance is one minus confidence. Also don't forget to repeat the line above for the upper confidence limit by replacing the first minus sign with a plus sign.

$$= AVERAGE(data) - T.INV(alpha, (COUNT(data) - 1)) * STDEV.S(data) / SQRT(COUNT(data))$$

Again chose the sign to match which one-sided confidence limit you require, minus for lower and plus for upper.

Binary Data

To calculate a confidence interval for a proportion in binary data, known as either a binary confidence interval or a binomial confidence interval, the following information is required,

- The number of successes or equivalent metric of interest.

- The number of trials in the sample: the sample size.

- The confidence level required or the level of risk that is acceptable.

- The interval of interest: one-sided or two-sided bounds.

Example 5.4 shows not only how to manually calculate a binary confidence interval in R to understand the steps involved, but also shows the shortcut using the **Hmisc** package. This shortcut only works for two-sided intervals unless you adjust the confidence level to compensate.

Jump straight to "**# Load library**" if you are only interested in the shortcut.

EXAMPLE 5.4

Calculate a 90% two-sided confidence interval and a one-sided upper confidence bound for 20 successes out of 25 trials.

```
# Assign the successes, trials, and significance level
x = 20; n = 25; alpha = 0.10
```

```
# Assess the probability of success
x/n
```

```
[1] 0.8
```

```
# Assign degrees of freedom for the critical values
# from the exact binomial method
df1l = 2*(n - x + 1); df2l = 2*x
df1u = df2l + 2; df2u = df1l - 2

# Calculate the two-sided confidence interval
lci = ifelse(x > 0, x / (x + qf(1 - alpha/2, df1l, df2l) *
      (n - x + 1)), 0)
uci = ifelse(x < n, ((x + 1) * qf(1 - alpha/2, df1u, df2u)) /
      (n - x + (x + 1) * qf(1 - alpha/2, df1u, df2u)), 1)
lci; uci
```

```
[1] 0.6245949
[1] 0.9177091
```

```
# Calculate the one-sided upper bound - replace alpha/2 with alpha
uci1 = ifelse(x < n, ((x + 1) * qf(1 - alpha, df1u, df2u)) /
      (n - x + (x + 1) * qf(1 - alpha, df1u, df2u)), 1)
uci1
```

```
[1] 0.8993822
```

```
# Load library
library(Hmisc)
```

```
# Shortcut to calculate the two-sided interval
binconf(x = 20, n = 25, alpha = 0.1, method = "exact")
```

```
PointEst      Lower       Upper
     0.8  0.6245949  0.9177091
```

```
# Shortcut to calculate the one-sided interval with amendment
binconf(x = 20, n = 25, alpha = 0.2, method = "exact")
```

```
PointEst      Lower       Upper
     0.8  0.6603411  0.8993822
```

The two-sided confidence interval is (0.625, 0.918) or 62.5% to 91.8%. Note that the probability of success was 80% that would make the intervals minus 17.5% and plus 11.8%, which is not symmetrical due to the probability of success being bound by 100%. The one-sided upper bound is 0.899 or 89.9% which is closer to 80% due to the risk placement.

Note: If using the shortcut to calculate a one-sided interval you will need to double the significance level, here from 0.1 to 0.2, and make sure you only quote one side and not both.

Translating the output from a binary confidence interval is important as you need to remember the interval is around an estimate and not a proportion of the sample. Using Example 5.4 you could state that you are "90% confident that the true success rate lies within 62.5% and 91.8%." You cannot state the sample success rate ± a value as the confidence limits are not symmetrical due to the bounding of 0 and 1. For the one-sided intervals you could state that you are "90% confident that the true success rate is less than 89.9%."

As binary data has far less "detail" than continuous data, the result is 0 or 1 as opposed to being any decimal place between two values. A much larger sample size will be required for a more precise interval. As with the continuous data examples, the confidence level and the sample size will have a large effect on the precision of the interval.

Example 5.5 uses the same data as Example 5.4 but this time looks at using a lower confidence level of 70%, then using a much smaller sample size of 5 trials with 4 successes.

EXAMPLE 5.5

Calculate a 70% two-sided confidence interval from the values in the previous example, then a 90% two-sided interval for 4 successes out of 5 trials, which is a much smaller sample size.

```
# Calculate the confidence interval at 70% confidence
binconf(x = 20, n = 25, alpha = 0.3, method = "exact")
```

```
PointEst     Lower      Upper
     0.8  0.6838798  0.8856574
```

```
# Calculate the confidence interval for the smaller sample size
binconf(x = 4, n = 5, alpha = 0.1, method = "exact")
```

```
PointEst     Lower      Upper
     0.8  0.3425917  0.9897938
```

The original confidence interval width with a sample size of 25 and a confidence level of 90% was roughly 29.3% (91.8% - 62.5%), decreasing the confidence level to 70% reduces that width to roughly 20.2% (88.6% - 68.4%); decreasing the sample size to 5 with a confidence level of 90% (while maintaining the same probability of success) increases that width to 64.7% (99.0% - 34.3%).

There are many methods for calculating binary confidence intervals, but arguably the three most common ones are the asymptotic normal interval or Wald interval, the Wilson interval, and the exact method, also known as the Clopper–Pearson interval.

The asymptotic normal interval is a normal approximation for binary data, and as such has some strong assumptions. As it is based on the normal distribution the interval will be symmetrical; it could go above 1 and below 0. It should only be used with large sample sizes and when the probability of success is not very small or very large, otherwise the results will be misleading.

The Wilson interval is more robust to deviations from the normal distribution than the asymptotic normal interval, and the intervals are not symmetrical so will converge at 0 or 1. It is easier to compute than the exact method and may be preferable with a medium to large sample size.

The exact method, which was used in the earlier examples, is the recommended method to use with small or medium sample sizes. The equations however are complicated to explain and they can use either the *F* distribution or the beta distribution to calculate the binomial cumulative density function (cdf), as such that will be left to personal investigation if you are interested. This method also can cope if the probability of success is 0 or 1 however it will produce conservative results meaning the confidence intervals will be wider than the other methods.

If you need to calculate binary confidence intervals in Excel, use the following functions for the lower and upper bounds for the two-sided interval, this uses the exact method:

$$= IF(x = 0, IF(x = n, (alpha/2)^\wedge(1/n), BETA.INV(alpha/2, x, n - x + 1))).$$

$$=IF(x = n, 1, IF(x = 0, 1 - (alpha/2)^\wedge(1/n), BETA.INV(1 - alpha/2, x + 1, n - x))).$$

Where *x* is the number of successes, *n* is the number of samples/trials, and alpha is the significance level, which is one minus the confidence level.

For the one-sided intervals use the following functions again this uses the exact method:

$$= IF(x = 0, IF(x = n, alpha^\wedge(1/n), BETA.INV(alpha, x, n - x + 1))).$$

$$=IF(x = n, 1, IF(x = 0, 1 - (alpha)^\wedge(1/n), BETA.INV(1 - alpha, x + 1, n - x))).$$

Make sure you chose only one of the functions from the two above, either the top line for the one-sided lower bound or the lower for the one-sided upper bound.

Tolerance Intervals

Tolerance intervals are used when you are interested in the uncertainty around a proportion of the population, which is termed the *coverage*.

Although you would want a confidence interval around the journey time to a specific location, average experience; you may not want a confidence interval around flying hours. For example, if you were trying to restrict flying hours to deal with pilot fatigue, it is much more sensible to calculate a tolerance interval, individual experience.

A confidence interval would show their average flying hours with uncertainty, whereas a tolerance interval would show a percentage (i.e., 95%) of their flying hours with uncertainty, which gives you far more information than the average of 50%.

As with confidence intervals, tolerance intervals can be one-sided or two-sided and that means that you either quote one tolerance limit or both limits. The tolerance interval values are called the lower tolerance limit/bound and the upper tolerance limit/bound; so for a one-sided tolerance interval you would quote either the lower bound or the upper bound, whereas for a two-sided tolerance interval you would quote both bounds

Tolerance intervals still incorporate confidence, but this is around the coverage and not around a sample estimate such as the mean.

The next sections look at calculating tolerance intervals for continuous and binary data and the information concerning sides also applies to binary data.

Continuous Data

To calculate a tolerance interval for continuous data that approximately follows a normal distribution the following information is required:

- The mean of the sample.

- The variability of the sample: the standard deviation.

- The number of measurements in the sample: the sample size.

- The confidence level required: or the level of risk that is acceptable.

- The coverage level required: what percentage of the data is of interest.

- The interval of interest: one-sided or two-sided bounds.

Example 5.6 shows not only how to manually calculate a tolerance interval in R to understand the steps involved, but also shows the shortcut using the **tolerance** package.

Jump straight to "**# Load library**" if you are only interested in the shortcut.

EXAMPLE 5.6

Calculate a 75% coverage, 90% confidence two-sided tolerance interval and a 75% coverage, 90% one-sided lower tolerance bound for the first data set in the chapter, data7.

```
# Calculate the mean, standard deviation and sample size for the data
# Assign the coverage level and the confidence level
x = mean(data7); s = sd(data7); n = length(data7); P = 0.75
conf = 0.9; x; s; n; P; conf
```

```
[1] 27.36278
[1] 0.7585885
[1] 18
[1] 0.75
[1] 0.9
```

```
# Calculate the values needed in the equation, including critical values
# from the Gaussian distribution and the Chi-squared distribution
n2 = (n - 1)*(1 + 1/n)
ncrit = (qnorm((1 - P)/2))^2
ccrit = qchisq(1 - conf, n - 1)
```

```
# Calculate the two-sided k
k2 = sqrt((n2*ncrit)/ccrit); k2
```

```
[1] 1.53445
```

```
# Calculate the two-sided tolerance interval
lower.2s.TI = x - k2*s
upper.2s.TI = x + k2*s
lower.2s.TI; upper.2s.TI
```

```
[1] 26.19876
[1] 28.52679
```

```
# Calculate the values needed for the one-sided tolerance interval
ncritcov = qnorm(P)
ncp = sqrt(n) * ncritcov
tcrit = qt(conf, df = n - 1, ncp = ncp)
```

```
# Calculate the one-sided k
k1 = tcrit/sqrt(n); k1
```

```
[1] 1.070626
```

```
# Calculate the one-sided lower tolerance bound
lower.1s.TI = x - k1*s; lower.1s.TI
```

```
[1] 26.55061
```

```
# Load library
library(tolerance)
```

```
# Shortcut to calculate the two-sided interval
normtol.int(data7, alpha = 0.1, P = 0.75, side = 2, method = "HE2")
```

	alpha	P	x.bar	2-sided.lower	2-sided.upper
1	0.1	0.75	27.36278	26.19876	28.52679

```
# Shortcut to calculate the one-sided interval
normtol.int(data7, alpha = 0.1, P = 0.75, side = 1)
```

	alpha	P	x.bar	1-sided.lower	1-sided.upper
1	0.1	0.75	27.36278	26.55061	28.17494

The two-sided tolerance interval is (26.20, 28.53) or 27.36 ± 1.16 due to symmetry, and the one-sided lower bound is 26.55.

Note: You can see that these bounds are larger than those for the confidence intervals, (27.05, 27.67) and 27.12, respectively, due to tolerance intervals covering a proportion of the data and not just the mean.

As shown in Example 5.6 if both the lower and upper one-sided bound had been calculated and incorrectly quoted as a two-sided interval, the values would be the same as calculating an 80% two-sided confidence interval—as 10% of risk would be at each end of the distribution totaling 20% risk overall.

Translating the output from a tolerance interval is important as you need to remember the interval is around a coverage or proportion. Using Example 5.6 you could state that you are "90% confident that at least 75% of the popula-tion lies within 26.20 and 28.53," or you could say that you are "90% confident that at least 70% of the population will be greater than 26.55" using an appro-priate number of decimal points dependent on required degree of accuracy.

As before, the confidence level and the sample size will have a large effect on the precision of the interval. In addition to this, the larger the coverage required the larger the interval will be.

Example 5.7 uses the same data as Example 5.6, data7, but this time looks at using a lower confidence level of 70% with the same coverage as before, then using a higher coverage of 95% with the same confidence level as before to highlight the previous points. The sample size will remain the same as the previous examples have already shown the benefits of increasing the sample size.

EXAMPLE 5.7

Calculate a 75% coverage, 70% confidence two-sided tolerance interval, then a 95% coverage, 90% confidence two-sided tolerance interval for the first data set, data7.

```
# Calculate the two-sided tolerance interval at 75% coverage,
# 70% confidence
normtol.int(data7, alpha = 0.3, P = 0.75, side = 2, method = "HE2")
```

	alpha	P	x.bar	1-sided.lower	1-sided.upper
1	0.3	0.75	27.36278	26.35784	28.36772

```
# Calculate the two-sided tolerance interval at 95% coverage,
# 90% confidence
normtol.int(data7, alpha = 0.1, P = 0.95, side = 2, method = "HE2")
```

	alpha	P	x.bar	1-sided.lower	1-sided.upper
1	0.1	0.95	27.36278	25.37953	29.34603

The original tolerance interval with a coverage of 75% and a confidence level of 90% was ±1.16, decreasing the confidence level to 70% while maintaining 75% coverage reduces that to ±1.00; increasing the coverage to 95% while maintaining 90% confidence increases that to ±1.98.

The above calculations can only be used to determine tolerance intervals if:

- The data can be assumed to follow a normal distribution.

If this doesn't apply, one option is to use nonparametric tolerance intervals. However they can be quite complicated: They can either be very large due to being distribution free, or they will just be bound by the minimum and maximum values of your sample.

A better option to try first would be to transform the data using one of the transformations listed in Chapter 4. The key thing to remember when doing this is to back transform the information at the end to give the correct confidence interval. It is also worth noting that the interval will no longer be symmetrical. There is no example included but the data in Example 5.3 (data9 and data10) would be suitable to try for yourself.

If you need to calculate tolerance intervals in Excel there are aids online, however I would recommend using R for its simplicity, as you've seen above using the shortcut.

The key difference between confidence intervals and tolerance intervals is that tolerance intervals will always be wider due to the fact that a tolerance interval accounts for a proportion of data whereas a confidence interval only accounts for an estimate such as the mean.

Figure 5-2 shows similar summer temperatures data to that in Chapter 3, with both confidence and tolerance intervals marked, each location had 20 samples.

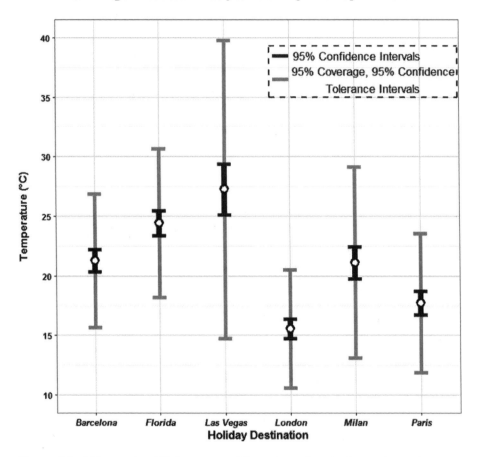

Figure 5-2. Difference in width between confidence intervals and tolerance intervals

Here we are 95% confident that the true mean lies within the red bounds for each location, and we are 95% confident that 95% of the population data lies within the green bounds. Note that these statements rely on the sample data being representative of the population that should have been satisfied by creating a good experimental design.

Binary Data

To calculate a tolerance interval for binary data, such as for acceptable or defective batches, the following information is required:

- The number of successes or equivalent metric of interest.

- The number of trials in the sample: the sample size.

- The confidence level required: or the level of risk that is acceptable.

- The coverage level required: what percentage of the data is of interest.

- The interval of interest: one-sided or two-sided bounds.

Example 5.8 shows not only how to manually calculate a tolerance interval in R to understand the steps involved, but also shows the shortcut using the **tolerance** package.

Jump straight to "**# Load library**" if you are only interested in the shortcut.

EXAMPLE 5.8

Calculate a 75% coverage, 90% confidence two-sided tolerance interval and a 75% coverage, 90% confidence one-sided lower tolerance bound for 20 successes out of 25 trials.

```
# Assign the successes, trials, coverage level, and significance level
x = 20; n = 25; P = 0.75; alpha = 0.10

# Assign the significance level and the coverage level for the
# two-sided interval
alpha = alpha/2; P = (P + 1)/2

# Calculate the two-sided tolerance interval using the exact method
lower.p = (1 + ((n - x + 1) * qf(1 - alpha, df1 = 2 * (n - x + 1),
          df2 = (2 * x)))/x)^(-1)
upper.p = (1 + (n - x)/((x + 1) * qf(1 - alpha, df1 = 2 * (x + 1),
          df2 = 2 * (n - x))))^(-1)
lower.p = max(0, lower.p); upper.p = min(upper.p, 1)
lower = qbinom(1 - P, size = n, prob = lower.p)
upper = qbinom(P, size = n, prob = upper.p)
lower; upper

[1] 13
[1] 24

# Reassign the successes, trials, coverage level and significance level
x = 20; n = 25; P = 0.75; alpha = 0.10

# Calculate the one-sided lower bound using the exact method
# Note same equation as previous with alpha left at 0.10
# Same would apply to an upper bound
lower.p = (1 + ((n - x + 1) * qf(1 - alpha, df1 = 2 * (n - x + 1),
          df2 = (2 * x)))/x)^(-1)
lower.p = max(0, lower.p)
lower = qbinom(1 - P, size = n, prob = lower.p)
lower
```

```
[1] 15
```

```
# Load library
library(tolerance)
```

```
# Shortcut to calculate the two-sided interval
bintol.int(x = 20, n = 25, P = 0.75, alpha = 0.1, side = 2,
    method = "CP")
```

```
    alpha    P      p.hat  2-sided.lower  2-sided.upper
1   0.1   0.75      0.8           13             24
```

```
# Shortcut to calculate the one-sided interval
bintol.int(x = 20, n = 25, P = 0.75, alpha = 0.1, side = 1,
    method = "CP")
```

```
    alpha    P      p.hat  1-sided.lower  1-sided.upper
1   0.1   0.75      0.8           15             24
```

The two-sided tolerance interval is (13, 24), which would be roughly 52% to 96%. Note that the number of successes was 20 and that would make the intervals -7 and +4, which is not symmetrical due to the successes being bound by 0 and 1. The one-sided lower bound is 15 or roughly 60%, which is closer to 20 due to the risk placement.

Note: You can see that these two-sided bounds are larger than those for the confidence intervals, (62.5, 91.8), due to tolerance intervals covering a proportion of the data and not just the mean.

Translating the output from a binary tolerance interval is important as you need to remember the interval is around a coverage or proportion. Using Example 5.8 you could state that you are "90% confident that at least 75% of future acceptable/defective batches lie within 13 and 24," you could state it as between 52% and 96%, however these figures have been rounded due to finding whole numbers in the calculation. You cannot state the sample successes plus/minus a value as the tolerance limits are not symmetrical due to the bounding of 0 and 1. For the one-sided interval you could say you are "90% confident that at least 75% of future acceptable/defective batches will be greater than 15."

Note that these calculations should only be used when future experiments will be using the same or a larger sample size, otherwise the results will not be correct.

As binary data has far less "detail" than continuous data, the result is 0 or 1 as opposed to being any decimal place between two values; a much larger sample size will be required for a more precise interval. As before, the confidence level, sample size, and coverage will have a large effect on the precision of the interval.

Example 5.9 uses the same data as Example 5.8 but this time looks at using a lower confidence level of 70% with the same coverage as before, then using a larger coverage of 95% with the same confidence level as before to highlight the earlier points. The sample size will remain the same as the previous examples have already shown the benefits of increasing the sample size.

EXAMPLE 5.9

Calculate a 75% coverage, 70% confidence two-sided tolerance interval, then a 95% coverage, 90% confidence two-sided tolerance interval for 20 successes out of 25 trials.

```
# Calculate the two-sided tolerance interval at 75% coverage,
# 70% confidence
bintol.int(x = 20, n = 25, P = 0.75, alpha = 0.3, side = 2,
    method = "CP")
```

	alpha	P	p.hat	2-sided.lower	2-sided.upper
1	0.3	0.75	0.8	14	24

```
# Calculate the two-sided tolerance interval at 95% coverage, 90%
confidence
bintol.int(x = 20, n = 25, P = 0.95, alpha = 0.1, side = 2,
    method = "CP")
```

	alpha	P	p.hat	2-sided.lower	2-sided.upper
1	0.1	0.95	0.8	11	25

The original tolerance interval width with a coverage of 75% and a confidence level of 90% was roughly 11 (24 - 13), decreasing the confidence level to 70% while maintaining 75% coverage reduces that width to roughly 10 (24 - 14); increasing the coverage to 95% while maintaining 90% confidence increases that width to roughly 14 (25 - 11).

There are many methods for calculating binary confidence intervals that I won't discuss, but the method used here is the exact method as with the confidence intervals. Although the interval is more conservative, it is more reliable with small or medium sample sizes.

It would get quite fiddly to calculate binary tolerance intervals in Excel, so again I would recommend using R for simplicity.

Figure 5-3 shows the number of detections per piece of equipment on the left and the same data as detection rate per piece of equipment on the right with both confidence and tolerance intervals marked. Depending on the emphasis of your study, you may want values (defective units) or proportions (detection rate). The sample size associated with each piece of equipment was 15, 13, 14, and 11, respectively, always quote the sample size with your results.

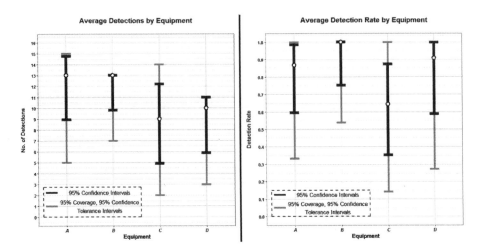

Figure 5-3. Difference in width between confidence intervals and tolerance intervals

Here we are 95% confident that the true success rate lies within the red bounds for each location, and we are 95% confident that 95% of the population data lies within the green bounds. Note that these statements rely on the sample data being representative of the population that should have been satisfied by creating a good experimental design.

Prediction Intervals

Prediction intervals are used when you are interested in the uncertainty around a future measurement.

For example, if you have measured the yield at certain concentrations, such as 3, 4, and 5, you may want to know the uncertainty around the next concentration at 6 or around a concentration you haven't recorded such as 4.5.

Prediction intervals are wider than confidence intervals as there is more uncertainty around a single value compared to a mean, which is made up of multiple values. Tolerance intervals can be larger or smaller than prediction intervals depending on how much coverage of the data is required.

As with confidence intervals and tolerance intervals, prediction intervals can be one-sided or two-sided, so make sure only one limit is quoted with one-sided intervals. The prediction interval values are called the lower prediction limit/bound and the upper prediction limit/bound.

Prediction intervals are generally used once a model has been fitted. We won't cover that until Chapter 7, so for now I will leave prediction intervals as just a description of what they are and deal with the calculations later.

Figure 5-4 shows an example of data that has been fitted to a model with the linear model or line of best fit, confidence intervals, tolerance intervals, and prediction intervals included to highlight the differences in interval width.

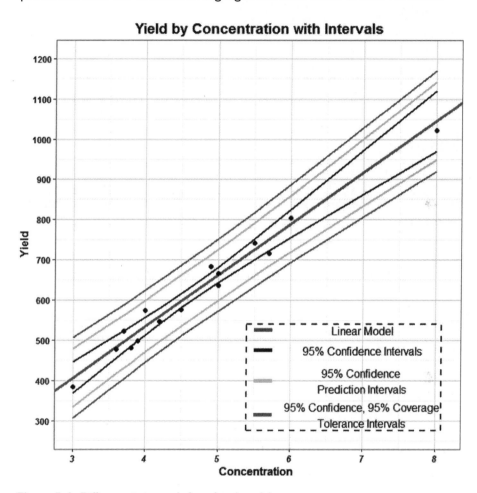

Figure 5-4. Difference in intervals for a fitted model

In this data there were 12 data points recorded, and then 3 extra points were used for predictions; at concentrations of 3, 4.5, and 8 with two of these being outside the originally recorded range.

The confidence intervals were calculated at the 95% confidence level. You can see how the lines approach the linear model, or line of best fit, where there is a lot of data and start to expand out at the extremes.

The prediction intervals also were calculated at the 95% confidence level; these aren't very reliable for data points outside the recorded range. You may be fine quoting the range at the concentration of 3 as it's quite close to the originally recorded data, however it wouldn't be advisable to make any assumptions about the concentration of 8 as it's far away from all the other sample data.

The tolerance intervals were calculated at 95% coverage and 95% confidence, and as such are larger than the prediction intervals, but they may not always be so. For example, if the tolerance intervals were calculated at 75% coverage and 95% confidence, in this example they would have been smaller than the prediction intervals.

Summary

The beginning of the chapter described what uncertainty is and the three most common types of intervals; confidence, tolerance, and prediction. It then went on to investigate these different intervals for both continuous data and binary data.

The confidence intervals section described that they represent uncertainty around a mean. It went into detail about intervals being either one-sided or two-sided, which holds for all other intervals. The section discussed the change in width of the intervals by altering either the confidence level or the sample size. It included worked examples for calculating continuous and binary confidence intervals both long-hand and using shortcuts in R.

The tolerance intervals section showed that they represent uncertainty around a coverage of the population with associated confidence. It demonstrated how changing the confidence level or the coverage level will affect the tolerance interval width. It compared the size of tolerance intervals to confidence intervals. The section also included examples for calculating continuous and binary confidence intervals.

The prediction intervals section was short due to the fact that these are calculated on data fitted to a model, which is not covered until Chapter 7. However, it explained that prediction intervals represent uncertainty around a value, such as a future measurement. The section did show how the size of these intervals compares to that of confidence intervals and tolerance intervals.

Chapter 6 moves on to "simple" hypothesis testing, with "*simple*" meaning only one or two variables are involved. Hypothesis testing forms the basis for most statistical testing; it involves comparing data to investigate for significant differences and the size of this difference or effect. The ideas of one-sided and two-sided will be carried across into hypothesis testing, and all output results in Chapter 6 includes confidence intervals you now are familiar with.

Hypothesis Testing

What Differences Are in the Data?

Hypothesis testing involves looking for significant differences within the sample of data you have, with *significant* being the important word here. It is easy to see if there has been a difference in results, between two sets of data for example, however hypothesis testing is used to determine if the difference is likely to be due to chance or likely to be a real effect in the population.

The idea of what hypothesis testing is doing is quite simple to understand, however the exact definition, the number of tests available for use, and the correct and incorrect language associated can be quite complicated. As such I always go along the lines of "you can think about the language incorrectly in your head for ease of understanding, but you must report it properly."

Hypothesis Test Components

No matter which hypothesis test method you use, the components will be the same in each. You will always require the data type and distribution to ascertain the correct method to use; you need to know what the actual hypotheses are; whether to conduct a one-sided or two-sided test, which would have been stated in the experimental design; and you also need to

©Victoria Cox 2017
V. Cox, *Translating Statistics to Make Decisions*, DOI 10.1007/978-1-4842-2256-0_6

know the corresponding level of confidence/significance that you are working toward. In addition the majority of tests will output a p-value, along with confidence intervals and a summary statistic such as the mean.

The second half of this chapter deals with the different hypothesis test methods that are available, and this section looks at the other components listed in the preceding paragraph.

Hypotheses

A hypothesis test has two hypotheses, the null hypothesis and the alternative hypothesis. The null hypothesis is what you are looking for evidence against. This hypothesis is generally the case where no action would need to be taken. The alternative hypothesis is what you are looking for evidence for, in an indirect way as this only occurs if there is evidence against the null, which isn't quite the same thing (see the "backward" p-value definition in the P-values section). The alternative hypothesis is generally the case where action would need to be taken.

For example, testing the current detector against a new detector, the customer wants to know if the new detector is significantly better than our current detector (see the Significant Differences section for more explanation on using the term *significant*). The null hypothesis here is that either there is no difference between the detectors, or the new detector performs worse than the current detector that is, if the new detector performs the same or worse than the current detector we will stick with the current detector, hence no action taken. The alternative hypothesis is that the new detector performs better than the current detector therefore, if the new detector performs better than the current detector we will need to purchase the new detector, hence action taken.

When dealing with the results you can simply think that if the p-value is big, the null hypothesis is likely and there is no significant difference. If the p-value is small, the alternative hypothesis is likely and there is a significant difference. However, do not ever quote the results as I've described them in that sentence as that is just an easy way to think about it in your head. As I've said previously when reporting results you will need to tweak the language to be correct.

Nothing is ever true or accepted in hypothesis testing, there is always "evidence against." For example, you wouldn't "accept the null hypothesis," there is just "no evidence to reject the null hypothesis," which suggests no evidence of a significant difference. In addition you wouldn't "accept the alternative hypothesis," there is just "evidence to reject the null hypothesis," which suggests evidence of a significant difference.

In terms of big and small p-values, that is only a general rule of thumb to remember which way indicates a significant difference. However, the value of the p-value that represents a significant difference depends on set confidence levels, which is discussed in the P-values and Significant Differences sections.

Sides or Tails

As with the intervals in Chapter 5 the side, or tail, of the test should have been decided during the experimental design phase, as it will have affected the required sample size.

A two-sided or two-tailed test means that you are looking for a significant difference on either side. For example, you want to know if there is a difference between the protection levels of Kit A and Kit B, so testing whether Kit A is better than Kit B and also whether Kit B is better than Kit A.

A one-sided or one-tailed test means you are looking for a significant difference at one particular side. For example, you want to know if Kit A is significantly better than Kit B, but you aren't concerned with the reverse scenario, as with the detector example in the Hypotheses section above.

It doesn't matter which hypothesis test method you decide you need to use due to the data you have collected, all hypothesis test methods will have the option to run a one-sided or two-sided test.

Similar to the uncertainty intervals, running a one-sided hypothesis test in both directions and quoting the results would be incorrect. If you are interested in a difference regardless of direction, then a two-sided test needs to be conducted.

P-values

When conducting hypotheses tests, the majority of them will give output that includes a p-value. There is a common misconception that a p-value represents the probability of the null being correct, which isn't strictly true. Given that the null hypothesis is true, the p-value shows how likely it is that you will see a sample such as the one you have found, hence my statement about the definition being "backward."

For example, if you were testing for a significant difference between two materials and the output showed a p-value of 0.003; and then if in reality there was no difference between the two materials, there is a 0.3% likelihood of seeing a sample as extreme as the one we have just collected. In other words, if we assume there is no difference between the materials in the population, it is very unlikely we would obtain a sample such as we have, therefore we can state evidence of a significant difference between the materials. A key thing to note here however is that the p-value is quoted on the assumption of "if the null hypothesis is true."

Most people won't state p-values and significant differences in terms of the proper definition, they will just state that there is evidence of a significant difference at the x% confidence level. This is fine as long as it is understood what the p-value actually represents and therefore what conclusions can be drawn from the results.

Significant Differences

To determine whether a p-value represents a significant difference it needs to be compared to a significance level, which would have been decided during the experimental design stage. Do not fall into the trap of just using 95% confidence/5% significance, just because that is the most widely used value. Refer back to Chapter 1 to remind yourself about confidence and power levels and what they mean in terms of risk.

Say the customer had decided to work at 95% confidence, which is 5% significance, and the p-value came out as 0.02 from the hypothesis test. Then we could state evidence of a significant difference, as 0.02 is smaller than 0.05. However, if the p-value had been 0.08 then we could not state evidence of a significant difference, as 0.08 is larger than 0.05. It may be worth noting there was evidence of a significant difference at 90% confidence, or just quote the p-value. On the other hand if the p-value had been 0.006 it would be worth stating that there was evidence of a significant difference at the 99% confidence level, not just at the 95% confidence level that we were working toward.

Significant differences also can be stated in terms of significance levels, for example saying there is evidence of a significant difference at the 90% confidence level is the equivalent of saying there is evidence of a significant difference at the 10% significance level.

It's worth getting into the habit of saying "there was evidence of a significant difference" rather than just "there was a significant difference" due to the uncertainty involved in the results including the big assumption about the null hypothesis being true.

To add even more translation options, significant differences also can be quoted in terms of strength of evidence:

- Strong evidence of a significant difference: is when the p-value < 0.01.

- Evidence of a significant difference: is when the p-value < 0.05.

- Weak evidence of a significant difference: is when the p-value < 0.1.

However not everyone is aware of this general rule of thumb, so it is preferable to state significant differences in relation to confidence or significance levels.

Make sure the sample size is quoted with all significance statements or p-values as this can help with interpretation of the results. For example, seeing a nonsignificant p-value with a sample size of 50 has much more weight for decision making than seeing the same nonsignificant p-value with a sample size of 3 as there is clearly not enough data to make an evidence-based decision.

Practical Differences

In all statistical testing the practical application needs to be considered at the forefront. It may be that evidence of a significant difference was found during the testing, however in practical terms that result is meaningless. The one thing that experimental design and general planning can't account for is the variation of the data. It may be that an adequate sample size was used, but the variation was much smaller than expected, and therefore a smaller difference between the two groups was classed as significant.

For example, an exaggerated example, if a study was conducted with a large sample size and the results showed that women ran a race in an average of 15 minutes and men in an average of 16 minutes, but both only had a standard deviation of 30 seconds and this resulted in evidence of a significant difference, this would not be practical. Obtaining a significant difference of 1 minute, in this example, would be meaningless and as such could be quoted as follows, there was evidence of a significant difference between the genders, however this difference was on average by 1 minute and therefore is not a practical difference or is not a difference of scientific interest.

The converse also could be true, if there was a study being conducted to determine whether future funding should be given to a new research area, then while evidence of a significant difference may not have been found per-haps due to cost restrictions on sample sizes, a practical difference may be of interest to the decision makers regardless.

There is always a balance between the output of a hypothesis test and the practicality of the size of the effect.

Plots

As mentioned in Chapter 3, box plots are a very useful tool for highlighting significant differences. However care needs to be taken when using them. They should only be used to highlight results found from a hypothesis test as the plots themselves cannot be used to determine significant differences, see Figure 6-1.

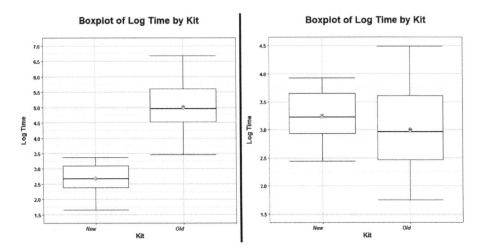

Figure 6-1. Example box plots of nonoverlapping and overlapping groups

Just because two boxes on a box plot do not overlap does not necessarily mean there will be a significant difference between the groups, which is a common misconception. For example, it may be due to a small sample size, so while there may look like a difference on the plot, there is not enough data for the test to determine that it's not just due to chance that the sample looks like it does. Although it is more likely that if the boxes largely overlap there is unlikely to be a significant difference, it still needs to be tested.

You cannot use a plot to make significance statements, a hypothesis test always needs to be carried out; this is why sample sizes should be quoted along with p-values of significance statements.

Interpretation

In terms of interpreting the output of a hypothesis test, as long as you follow a step-by-step pattern, until you are more familiar with the output, you should get the correct answer:

- Run the correct test for the data collected and obtain a p-value—see the Hypothesis Tests section for the different methods available.

- Compare the p-value to the predetermined significance level.

- Decide whether that means you have evidence to reject the null hypothesis.

- Turn that statement back into English while still thinking about null and alternative hypotheses.

- Answer the initial question posed.

For example, consider the question "is there a difference between Machine A and Machine B in terms of average computation time." Having run a two-sample *t* test and obtaining a p-value of 0.02 and working to 95% confidence, we now follow the remaining steps:

- p-value of 0.02 is less than the significance level of 0.05.

- Therefore there is evidence to reject the null hypothesis.

- Therefore there is evidence to reject that the two machines have the same computation time.

- There is evidence of a significant difference between the computation times of the machines at the 95% confidence level, with Machine A being quicker on average by 10 minutes.

It is clear that without data that last part of the example was made up, but the principal holds that in addition to stating evidence of a significant difference, you also need to qualify that in terms of direction and size of the difference.

Hypothesis Tests

Now that you have an idea of the components that make up a hypothesis test, what they mean, and what the output will include, it's time to look at some of the tests available to use.

The tests in this chapter are limited to one or two sample groups only as Chapter 7 deals with larger data sets though the interpretation of the output will be roughly the same as with these tests.

The type of test to use depends on three things, the type of data you have, the number of samples, and the relationship between multiple samples. So you need to know if you have one sample of data or two samples, then within two samples whether the data is independent or paired. In terms of data type there will be continuous or binary, then within continuous data whether it is normally distributed or not. You also should remember that the continuous data tests are conducted on the averages.

Figure 6-2 shows which hypothesis tests should be used dependent on the data you have collected.

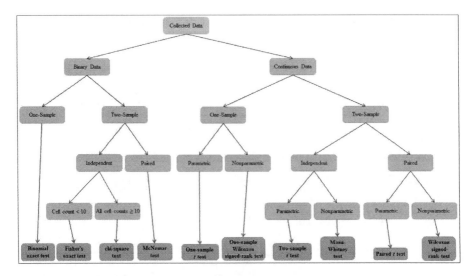

Figure 6-2. Flowchart that shows recommended hypothesis tests depending on the data

It is worth noting though that the output from all the tests will look similar, the difficulty is initially choosing the correct test to use and then interpreting the output correctly, again there will be some similarities with this as well.

Another item to be aware of is conducting multiple tests and quoting significant differences for each one independently. For example, if you had a continuous normally distributed data set with 4 groups and you wanted to test for significant differences between each one, it would require 6 tests. These tests would not be independent and corrections for running multiple tests will need to be applied, such as Bonferroni corrections (more on this in Chapter 7).

Binary Data

With binary data there are only three possible options to choose between, one-sample, two-sample independent, or paired.

One-Sample Binary Data

With binary data that has one sample of data to be compared to a value, such as a threshold limit, the hypothesis test to use is an exact binomial test.

It's worth noting that there is an alternative proportions test that could be used, either with or without Yates's continuity correction dependent on sample size. However, this is only an approximation; it is a simpler calculation than that of the exact binomial test. As calculations no longer need to be calculated by hand, the exact binomial test is recommended as it is preferable with a smaller sample size anyway, and even if the sample size is large the p-value will be roughly equal to that of the proportions test.

Setting up an exact binomial test requires the number of "successes"; the number of trials; a null hypothesis that will be a threshold value or equivalent; an alternative hypothesis that will take the form *"less," "greater,"* or *"two. sided"* in R; and a confidence level.

Example 6.1 shows how to run a one-sample exact binomial test using protection levels.

EXAMPLE 6.1

The customer wants to know if our equipment is providing greater than 70% protection for our troops at 95% confidence. Note here that to "prove" they are protected our null hypothesis will assume that they aren't protected. We had 30 participants and 26 of them were protected in the trial.

```
# Run the one-sample binomial test
binom.test(x = 26, n = 30, p = 0.7, alternative = "greater",
    conf.level = 0.95)
```

Exact binomial test

data: 26 and 30
number of successes = 26, number of trials = 30, p-value = 0.03015
alternative hypothesis: true probability of success is greater than 0.7
95 percent confidence interval:
0.7203848   1.0000000
sample estimates:
probability of success
    0.8666667

The p-value is 0.030, which is smaller than our significance level of 0.05, which suggests evidence to reject the null hypothesis. In terms of answering the question we can say that there is significant evidence that the troops are protected.

In addition the output shows the sample "probability of success," which is 86.7% and it also shows a 95% lower confidence limit, which is 72.0%. This highlights how to interpret the results as even the lower confidence limit for the mean protection is above our threshold limit of 70%.

In Example 6.1 the null hypothesis is that the protection level is 70% or less, which makes the alternative hypothesis that the protection level is above 70% as in this case we want to assume the worst scenario and "prove" the best scenario. This is why in the R code the alternative is chosen as "greater."

The conclusion from the example is that there was evidence that the protection of our troops is significantly higher than the threshold of 70% at the 95% confidence level. Therefore our troops are sufficiently protected as the lower bound for protection is 72.0%.

Two-Sample Binary Data

With binary data that has two samples of data to be compared against each other, the hypothesis test to use is either a Fisher's exact test or a chi-square test with Yates's continuity correction. The chi-square test doesn't provide confidence intervals on the results, so the best way to conduct that test is to run a proportions test with continuity correction.

The first thing to do is create a contingency table, a 2 × 2 table, of the counts. A general rule of thumb is that if any of these counts are below 10 then you need to use Fisher's exact test, but if all the cell counts are 10 or above the Fisher's is still best, but you can use the chi-square test if you prefer. With a large sample size the Fisher's exact test will give similar results to the chi-square test.

Generally binary data is recorded as 1 for the results of interest and 0 for those that aren't, however to aid with the examples the results will be labeled with text.

Example 6.2 shows how to run both a Fisher's exact test and a chi-square test. In addition there is a shortcut to creating row percentages that can be run using the **RcmdrMisc** package.

EXAMPLE 6.2

The customer wants to know if there is a difference between the two treatments A and B at 99% confidence. Note here that subjects couldn't have both treatments, hence independence. We had 30 subjects for each treatment and the count of those cured was 25 and 16 for each treatment, respectively.

```
# Set up the data
results = matrix(c(25, 5, 16, 14), ncol = 2, byrow = TRUE)
rownames(results) = c("A", "B")
colnames(results) = c("Cured", "X")

# Assign the data to a table and print results
Table = as.table(results)
Table

   Cured   X
A    25    5
B    16   14
```

As one of the cell values in the contingency table was below 10, Fisher's exact test should be used. Note if all cell values had been 10 or above you could use the chi-square test.

```
# Load library and create row percentages
library(RcmdrMisc)
rowPercents(Table)
```

```
   Cured     X  Total  Count
A  83.3   16.7    100     30
B  53.3   46.7    100     30
```

```
# Run the appropriate test - Fisher's exact test
fisher.test(Table, conf.level = 0.99, alternative = "two.sided")
```

```
Fisher's Exact Test for Count Data

data:  Table
p-value = 0.02506
alternative hypothesis: true odds ratio is not equal to 1
99 percent confidence interval:
0.8322165    28.7131182
sample estimates:
odds ratio
 4.263791
```

The p-value is 0.025, which is not smaller than our significance level of 0.01, which suggests there is no evidence to reject the null hypothesis at 99% confidence. In terms of answering the question we can say that there is no evidence of a significant difference between the two treatments at the 99% confidence level. It may be worth noting there was evidence of a difference at 95% confidence, however due to the fact the data is from drug responses, the higher confidence level is probably a hard requirement.

Using a Fisher's exact test gives the odds ratio between the treatments, so here it is saying that the odds of being Cured using Treatment A is 4.26 times the odds of being Cured using Treatment B. However there is a confidence interval around this, so the odd ratio could be as small as 0.83 times or as large as 28.71.

```
# Create a plot of results
barplot(t(Table), main = "Outcome by Treatment", xlab = "Treatment",
        ylab = "Count", space = NULL, ylim = c(0,25), beside = TRUE,
        legend.text = TRUE, col = c("dodgerblue3","firebrick3"))
```

Outcome by Treatment

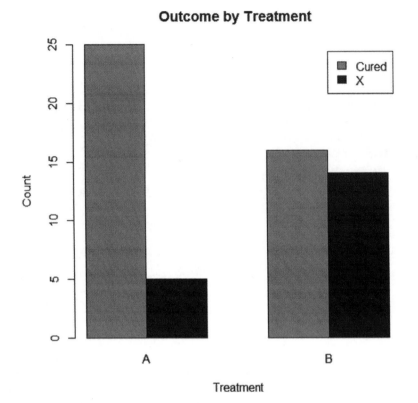

```
# Example code to show how to run the chi-square equivalent test
prop.test(Table, conf.level = 0.99)
```

2-sample test for equality of proportions with continuity correction

```
data:  Table
X-squared = 4.9294, df = 1, p-value = 0.0264
alternative hypothesis: two.sided
99 percent confidence interval:
-0.02618514   0.62618514
sample estimates:
   prop 1     prop 2
0.8333333  0.5333333
```

The code and output using the chi-square test also has been shown. In this example both outputs produced the same conclusions that there is no evidence of a significant difference between the Treatments at 99% confidence. However you can see that the chi-square test p-value is more conservative, 0.026 compared to 0.025, and this is due to it being an approximation rather than an exact calculation. The difference could be much larger in other studies so you need to make sure the correct calculation is chosen dependent on the contingency table cell frequencies.

The confidence interval here is around the difference between the two proportions of being Cured. The difference is 0.83 – 0.53, which is 0.30, however this could be as small as -0.03 or as large as 0.63. The negative value just means that in certain cases Treatment B may cure more than Treatment A.

In Example 6.2 one of the cell counts in the contingency table was 5, which is below 10 so Fisher's exact test is the appropriate test to use.

The customer just wanted to know if there was a difference between the treatments, which makes the null hypothesis that the treatments are equal and the alternative hypothesis that the treatments are different—hence the alternative being **"two.sided"** in the R code.

The conclusion from the example is that although Treatment A cured more subjects than Treatment B, 83.3% and 53.3%, respectively, there was no evidence of a significant difference between the treatments at the 99% confidence level.

Paired Binary Data

With binary data that has two samples of data that is paired, the hypothesis test to use is a McNemar test.

Again, as with the previous example, a contingency table will need to be created, then run the McNemar test on that table.

Example 6.3 shows an example of running a McNemar test using an example about hazmat suits. To run the test the R package **exact2x2** will be used, note this package does have dependencies. To plot the data the R package **ggplot2** and its dependencies will be required. Details about the structure of creating a ggplot is be discussed in Chapter 9.

EXAMPLE 6.3

The customer wants to know if there is a difference between two hazmat suits at 95% confidence. There were 32 subjects and each subject tried to complete a set course with both hazmat suits. They either completed the course in the hazmat suit, or had to remove it. The response data will be shown in the following example.

```
# Set up the data
results = matrix(c(8, 15, 5, 4), ncol = 2, byrow = TRUE)
rownames(results) = c("Suit 1 Complete", "Suit 1 Remove")
colnames(results) = c("Suit 2 Complete", "Suit 2 Remove")

# Assign the data to a table and print results
Table2 = as.table(results)
Table2
```

	Suit 2 Complete	Suit 2 Remove
Suit 1 Complete	8	15
Suit 1 Remove	5	4

```
# Load the library and run the McNemar test
library(exact2x2)
mcnemar.exact(Table2)
```

```
Exact McNemar test (with central confidence intervals)

data:  Table2
b = 15, c = 5, p-value = 0.04139
alternative hypothesis: true odds ratio is not equal to 1
95 percent confidence interval:
1.03647   10.55115
sample estimates:
odds ratio
    3
```

```
# Create plot of results
library(ggplot2)
data11 = as.data.frame(Table2)
ggplot(data11, aes(Var2, Var1)) + geom_tile(aes(fill = Freq),
       colour = "black") + scale_fill_gradient(low = "white",
       high = "steelblue") + theme_bw() + xlab("Suit 2") +
       ylab("Suit 1") +
       ggtitle("Frequency of Task Completion in Each Suit")
```

Frequency of Task Completion in Each Suit

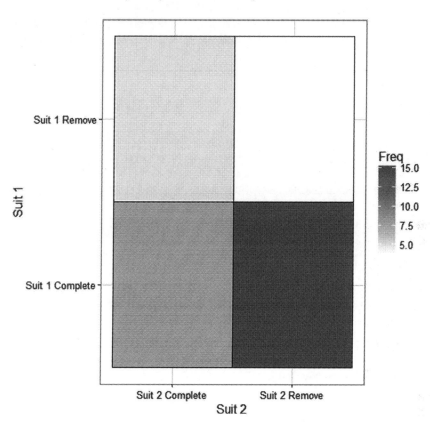

The p-value is 0.041, which is smaller than our significance level of 0.05, which suggests there is evidence to reject the null hypothesis at 95% confidence. In terms of answering the question we can say that there is evidence of a significant difference between the two hazmat suits at the 95% confidence level.

The next item to answer is how the two hazmat suits are different; if you look at the table of counts we can ignore the cases where the subjects completed the course with both suits and also where they removed both suits. The remaining values are 5 and 15, and if you look at the bottom of the McNemar test output you can see an odds ratio of 3, so the odds of completing the course is 3 times higher using Hazmat Suit 1 compared to using Suit 2 and there is a confidence interval around this odds ratio.

The McNemar test in Example 6.3 deals with odds ratios, so here the null hypothesis is that the odds ratio between the hazmat suits is 1, or that they are the same, and the alternative hypothesis is that the odds ratio between them is not equal to 1, or that they are different.

The conclusion from the example is that there was evidence of a significant difference between the two hazmat suits at the 95% confidence level, with the odds of completing the course in Hazmat Suit 1 being 3 times higher than the odds using Hazmat Suit 2. Therefore Hazmat Suit 1 is the preferable option.

Continuous Data

With averages of continuous data there are the same options as with binary data: one-sample, two-sample, or paired. In addition though, the distribution needs to be checked to determine whether parametric tests can be used with normally distributed data, or whether nonparametric tests need to be used with non-normally distributed data. However as mentioned in Chapter 4, always remember to try a transformation on the data first if the data appears non-normal or use the rule of thumb given for the confidence intervals for a mean concerning symmetry and sample size.

One-Sample Normally Distributed Data

For continuous data that has one sample of normally distributed data to be compared to a value, such as a threshold limit, the hypothesis test to use is a one-sample *t* test.

To assume normality you will need to visually check the distribution first—this requires the R package *car*, once that is confirmed then you can move on to conducting the one-sample *t* test. The test requires a null hypothesis, which will be the threshold value or equivalent an alternative hypothesis with the same options as those for the exact binomial test and a confidence level.

Example 6.4 shows an example for running a one-sample *t* test on quality control data.

EXAMPLE 6.4

The customer wants to know if our equipment passes quality control at the 90% confidence level; the threshold limit for the average weight of the equipment is 33kg and the batch will fail if it rises above this value. We had 19 measurements and the raw data will be shown in the example below.

```
# Set up the data
data12 = c(32.5, 32.8, 35.7, 34.6, 34.8, 33.7, 32.9, 35.3, 33.7,
           32.3, 32.0, 32.9, 33.6, 33.4, 34.1, 32.8, 32.5, 32.5, 34.0)

# Load the library and check for normality
library(car)
qqPlot(data12, dist = "norm", main = "Q-Q Plot for Normality",
       xlab = "Norm Quantiles", ylab = "Weight (kg)")
```

Q-Q Plot for Normality

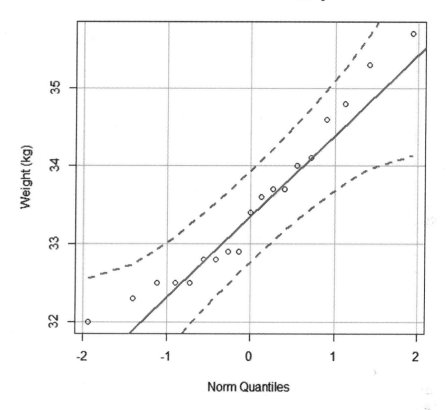

A quantile comparison plot shows the data to be approximately normally distributed—the line is roughly from the bottom left of the plot to the top right.

```
# Run the one-sample t test
t.test(data12, mu = 33, alternative = "greater", conf.level = 0.90)
```

```
One Sample t-test
```

```
data:  data12
t = 1.977, df = 18, p-value = 0.03178
alternative hypothesis: true mean is greater than 33
90 percent confidence interval:
33.15665      Inf
sample estimates:
mean of x
33.47895
```

```
# Create plot of results with the threshold limit and mean
boxplot(data12, xlab = "Batch of Equipment", ylab = "Weight (kg)",
        ylim = c(32,36), main = "Weight of Equipment")
abline(h = 33, lty = "dashed", col = "red")
points(mean(data12), pch = 19, col = "blue")
```

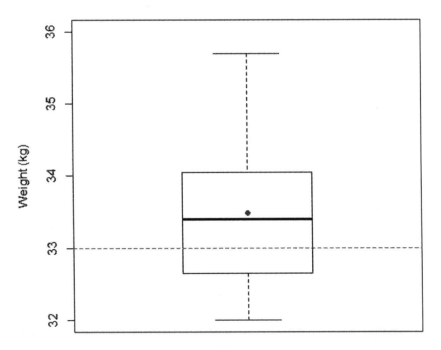

Weight of Equipment

Batch of Equipment

The p-value is 0.032, which is smaller than our significance level of 0.10, which suggests there is evidence to reject the null hypothesis at 90% confidence. In terms of answering the question we can say that there is significant evidence that the batch has failed.

In addition the output shows the sample mean, which is 33.48kg and it also shows the 90% lower confidence limit, which is 33.16kg. This highlights how to interpret the results as even the lower confidence limit for the mean weight is above the threshold limit of 33kg—also marked on the plot.

Example 6.4 states that the equipment will fail quality control if the average weight is significantly greater than 33kg, so this is the case where we would take action. This makes the null hypothesis that the weight is 33kg or less and the alternative hypothesis that the weight is above 33kg—hence why alternative is **"greater"** in the R code.

The conclusion from the example is that there was evidence that the average weight of the equipment is significantly higher than the threshold limit of 33kg at the 95% confidence level so the batch has failed as the lower bound of weight of equipment is 33.16kg.

One-Sample Non-Normally Distributed Data

For continuous data that has one sample of non-normally distributed data to be compared to a value, such as a threshold limit, the hypothesis test to use is a one-sample Wilcoxon signed-rank test.

Once you have visually confirmed the data is non-normal and have tried a transformation on the data, you can proceed with the one-sample Wilcoxon signed-rank test. This test requires a null hypothesis, which will be the threshold value or equivalent; an alternative hypothesis; and a confidence level.

In addition to the list in the previous paragraph, the default code does not produce a summary statistic or confidence intervals, so we need to specify if we want those. The summary statistic is a pseudomedian, and this will not necessarily match the median of the raw data as the pseudomedian is calculated from the median of the mean of pairs of data points. It is also worth specifying that an exact p-value should be calculated but with sample sizes less than 50 it will default to this case anyway.

Example 6.5 looks at a one-sample Wilcoxon signed-rank test for UK temperature data.

EXAMPLE 6.5

The customer wants to know if the average weekly UK temperatures are less than 12.5°C at the 90% confidence level, as this will require purchasing new clothing. We will use the weekly UK temperature data from Chapter 4, shown here, which contained 50 measurements.

```
# Set up the data
data13 = c(9.246734, 7.399515, 10.747294, 3.569408, 4.337869,
           3.172818, 14.205624, 30.076914, 15.747489, 6.751340,
           8.868595, 9.067760, 8.168440, 7.499503, 7.377515,
           14.883616, 19.688646, 26.299868, 6.351835, 14.180845,
           8.291489, 6.923344, 8.540164, 11.488742, 22.694856,
           16.868368, 31.439693, 10.700027, 17.887367, 10.008738,
           10.678093, 13.064685, 24.202956, 12.361150, 12.772815,
           13.436628, 14.336022, 4.701801, 6.078979, 16.039244,
           13.830606, 11.857714, 11.927977, 4.661250, 28.652883,
           6.391380, 4.378959, 8.361308, 11.056678, 7.521961)

# Load the library and check for normality
library(car)
qqPlot(data13, dist = "norm", main = "Q-Q Plot for Normality",
    xlab = "Norm Quantiles", ylab = "Temperature (°C)")
```

Q-Q Plot for Normality

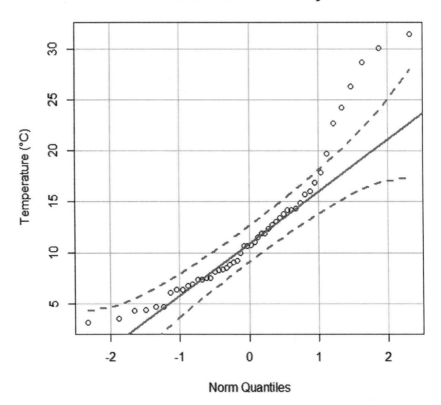

A quantile comparison plot shows the data is not normally distributed, the line is not straight from the bottom left of the plot to the top right, it is curved.

```
# Assume transformation didn't work
# Run the one-sample Wilcoxon signed-rank test
wilcox.test(data13, mu = 12.5, alternative = "less",
          conf.level = 0.90, conf.int = TRUE, exact = TRUE)
```

```
Wilcoxon signed rank test

data:  data13
V = 492, p-value = 0.08147
alternative hypothesis: true location is less than 12.5
90 percent confidence interval:
-Inf    12.44044
sample estimates:
(pseudo)median
    11.15571
```

```
# Create plot of results with the threshold limit
boxplot(data13, xlab = "Sample", ylab = "Temperature (°C) ",
        ylim = c(0,35), main = "Weekly UK Temperatures")
abline(h = 12.5, lty = "dashed", col = "red")
```

Weekly UK Temperatures

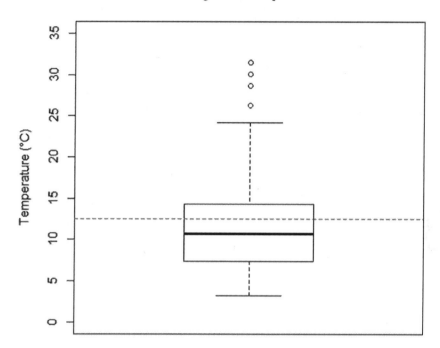

Sample

The p-value is 0.081, which is smaller than our significance level of 0.10 that suggests there is evidence to reject the null hypothesis at 90% confidence. In terms of answering the question we can say that there is significant evidence that the average weekly UK temperature is less than 12.5°C.

In addition the output shows the pseudomedian, which is 11.16°C and it also shows a 90% upper confidence limit, which is 12.44°C. This highlights how to interpret the results as even the upper confidence limit for the pseudomedian temperature is below the threshold limit of 12.5°C, which is marked on the plot.

In Example 6.5 the customer would need to take action and buy new clothing, if the average weekly UK temperature was above 12.5°C. This makes the null hypothesis that the "median" temperature is equal or greater than 12.5°C, and the alternative hypothesis that the "median" temperature is less than 12.5°C.

The conclusion from the example is that there was evidence that the median weekly UK temperatures are significantly lower than the threshold value of 12.5°C at the 90% confidence level. Therefore new clothing needs to be purchased as the upper bound of the average temperature is 12.44°C.

Two-Sample Normally Distributed Data

For continuous data that has two independent samples of normally distributed data with averages to be compared against each other, the hypothesis test to use is a two-sample *t* test.

After the normality check, as there are two samples you should check for equal variances. The default of the two-sample *t* test is to assume the variances are unequal, as this will provide a more conservative result, however if an equal variance test shows that you can assume equal variance, then you can amend the two-sample *t* test accordingly. The best way to check for equal variance is visually, using box plots, however there are also formal tests that can be used.

For data with two or more samples with an assumed normal distribution the equal variance test to use is called Bartlett's test. There is a two-variance *F* test for data with two samples and an assumed normal distribution, however Bartlett's test also covers this case so is the preferred method to use.

Once equal variances have been checked, you can move on to running the two-sample *t* test. This test requires an alternative hypothesis, a confidence level, and whether equal variances can be assumed.

Example 6.6 shows a two-sample *t* test on temperature data, it includes a formal equal variance test. However the box plot at the end of the example also can be used to assess the assumption of equal variances.

EXAMPLE 6.6

The customer wants to know if there is a difference between the average temperature readouts of two thermometers, new and old, at the 99% confidence level. We had 19 measurements for each thermometer and the raw data will be shown in the following example.

```
# Set up the data
Temp = c(21.6, 20.7, 22.8, 23.7, 22.4, 23.1, 20.9, 21.6, 22.2, 21.7,
         20.5, 23.4, 22.6, 22.4, 21.3, 20.6, 21.7, 21.9, 22.3, 22.9,
         23.6, 24.7, 25.1, 24.9, 23.7, 25.6, 24.7, 24.1, 23.1, 23.9,
         24.6, 25.2, 24.7, 24.3, 23.5, 23.8, 24.6, 25.0)
Type = rep(c("New", "Old"), each = 19)
data14 = data.frame(Temp, Type)

# Load library and check for normality
library(car)
qqPlot(data14$Temp, dist = "norm", main = "Q-Q Plot for Normality",
       xlab = "Norm Quantiles", ylab = " Temperature (Celsius)")
```

Q-Q Plot for Normality

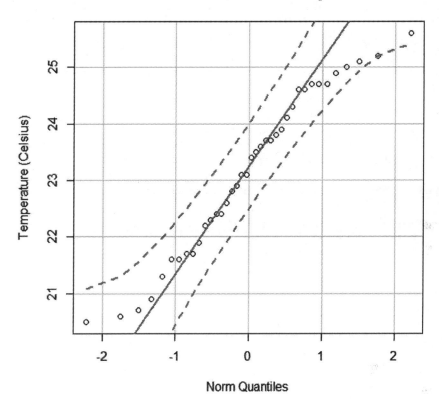

Norm Quantiles

A quantile comparison plot shows the data to be approximately normally distributed as the line is roughly from the bottom left of the plot to the top right.

```
# Check for equal variances
bartlett.test(Temp ~ Type, data = data14)
```

Bartlett test of homogeneity of variances

```
data:  Temp by Type
Bartlett's K-squared = 0.86468, df = 1, p-value = 0.3524
```

The Bartlett's test for equality of variance gave a p-value of 0.352, which is larger than the general significance level of 0.05, this suggests there is no evidence to reject the null hypothesis therefore we can assume equal variance. This test backs up the box plot that shows roughly similar shaped box and whiskers for each Thermometer type.

```
# Run the two-sample t test
t.test(Temp ~ Type, alternative = "two.sided", conf.level = 0.99,
       var.equal = TRUE, data = data14)
```

Two Sample t-test

```
data:  Temp by Type
```

```
t = -8.582, df = 36, p-value = 3.126e-10
alternative hypothesis: true difference in means is not equal to 0
99 percent confidence interval:
-3.091205   -1.603532
sample estimates:
mean in group New   mean in group Old
     21.96842            24.31579
```

```
# Create plot of results with the mean for each group
boxplot(Temp ~ Type, data = data14,
    main = "Temperature Readout by Thermometer",
    xlab = "Thermometer", ylab = "Temperature Readout (°C)")
points(mean(data14$Temp[data14$Type == "New"]), pch = 19,
    col = "blue")
points(mean(data14$Temp[data14$Type == "Old"]), x = 2, pch = 19,
    col = "blue")
```

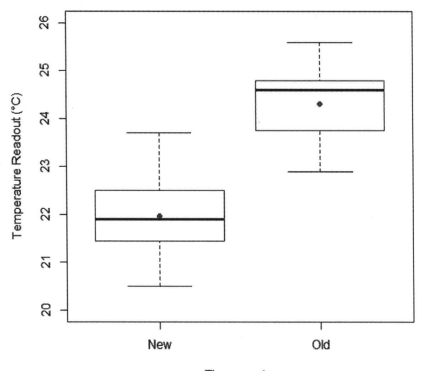

The p-value of the two-sample *t* test is 0.000, rounding to 3 significant figures, which is smaller than our significance level of 0.01, which suggests there is evidence to reject the null hypothesis at 99% confidence. In terms of answering the question we can say that there is evidence of a significant difference between the two thermometer readouts.

The output also shows the sample means for the two thermometers that shows the direction of the significant difference. They are 21.97°C and 24.32°C, respectively, which means that the new thermometer is showing significantly lower temperature readouts compared to the old thermometer. The average difference in temperature readout between them is 2.35°C, and note the confidence interval around this average difference is 1.60°C to 3.09°C.

Initially in Example 6.6 a box plot needed to be drawn to investigate equal variances as long as the size of the box and whiskers is similar then the variances can be assumed similar. A Bartlett's test also was run to confirm these findings, which is generally done at the 95% confidence level. The null hypothesis of this test is that the variances of the two samples are equal, and the alternative is that they are unequal.

The customer just wants to know if there is a difference between the temperature readouts, so this makes the null hypothesis that the readouts are the same, and the alternative hypothesis that the readouts are different—hence choosing alternative as *"two.sided"* in the R code.

The conclusion from the example is that there was evidence of a significant difference between the thermometers at the 99% confidence level, with the new thermometer showing lower temperature readouts, between 1.60°C to 3.09°C lower than the old thermometer.

Two-Sample Non-Normally Distributed Data

For continuous data that has two independent samples of non-normally distributed data to be compared against each other, the hypothesis test to use is a Mann–Whitney test, which also can be called the Wilcoxon rank-sum test.

After the normality check and confirming a transformation would not work, you need to check for equal variances. One of the assumptions of the Mann–Whitney test is that the samples have equal variance, so the test to use to establish whether this assumption is justified is called Levene's test.

Once equal variances have been checked, you can proceed with the Mann–Whitney test. This test uses the same code as the one-sample Wilcoxon signed-rank test with some amendments, so it requires two samples of data, an alternative hypothesis, and a confidence level.

The same extra specifications apply as with the previous use with regards to including a pseudomedian of the difference, confidence intervals, and using the exact calculation.

Example 6.7 looks at a Mann–Whitney test using distances fired for different projectiles. The package *car* is also used to conduct the Levene's test.

EXAMPLE 6.7

The customer wants to know if there is a difference in average distance fired between batches of projectiles made by two different companies at the 95% confidence level. We had 10 measurements for each companies' projectiles and the raw data is shown in the following example.

```
# Set up the data
A = c(80.1, 78.6, 70.9, 75.6, 77.4, 73.1, 65.7, 53.6, 52.8, 30.1)
B = c(31.8, 51.2, 49.8, 35.9, 71.7, 82.3, 80.2, 78.8, 46.7, 79.9)

# Load the library and check for normality
library(car)
qqPlot(A, dist = "norm", main = "Q-Q Plot for Normality",
       xlab = "Norm Quantiles",
       ylab = "Distance (m) for Projectiles A")
qqPlot(A, dist = "norm", main = "Q-Q Plot for Normality",
       xlab = "Norm Quantiles",
       ylab = "Distance (m) for Projectiles B")
```

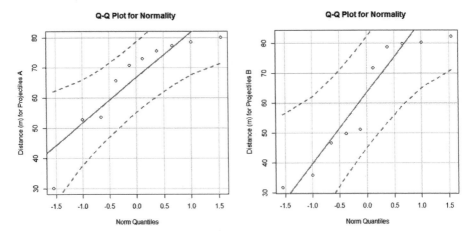

A quantile comparison plot shows both groups of data are not normally distributed. The lines are not straight from the bottom left of the plots to the top right; they are curved.

```
# Create a data frame
Dist = c(A, B)
Group = rep(c("A", "B"), each = 10)
data15 = data.frame(Dist, Group)

# Assume transformation didn't work and check for equal variances
leveneTest(Dist ~ Group, data = data15)
```

```
Levene's Test for Homogeneity of Variance (center = median)
      Df  F value  Pr(>F)
group  1    2.188  0.1564
      18
```

The spread is similar on the plots and Levene's test for equality of variance gave a p-value of 0.156, which is larger than the general significance level of 0.05 which suggests there is no evidence to reject the null hypothesis so we can assume equal variance.

```
# Run the Mann-Whitney test
wilcox.test(data15$Dist ~ data15$Group, alternative = "two.sided",
    conf.int = TRUE, conf.level = 0.95, exact = TRUE)
```

```
Wilcoxon rank sum test

data:   data15$Dist by data15$Group
W = 52, p-value = 0.9118
alternative hypothesis: true location shift is not equal to 0
95 percent confidence interval:
-9.3    26.4
sample estimates:
difference in location
        1.35
```

```
# Create plot of results
boxplot(Dist ~ Group, data = data15, ylim = c(20,90),
        main = "Distance Fired by Projectile", xlab = "Projectile",
        ylab = "Distance Fired (m)")
```

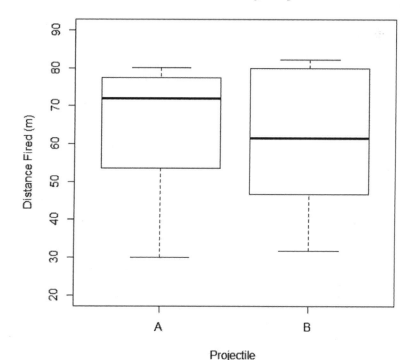

The p-value of the Mann–Whitney test is 0.912, which is larger than our significance level of 0.05, which suggests there is no evidence to reject the null hypothesis at 95% confidence. In terms of answering the question we can say that there is no evidence of a significant difference between the median distances fired for the two companies' projectiles.

The output also shows the sample difference in location between the two groups, which is 1.35 m, which means that the distance fired for projectile A is further than that of projectile B, but not significantly so. The confidence interval around this difference is -9.3m to 26.4m, which suggests large variation for a small sample size.

For Example 6.7, initially a Levene's test was run to investigate equal variances; this is generally done at the 95% confidence level. The null hypothesis is that the variances of the two samples are equal, and the alternative is that they are unequal and this was backed up by the box plot.

The customer wants to know if there is a difference between the distance fired of the projectiles, so this makes the null hypothesis that the distances are the same. The alternative hypothesis was that the distances are different, hence choosing alternative as **"two.sided"** in the R code.

The conclusion from the output is that there was no evidence of a significant difference between the companies' projectiles at the 95% confidence level.

Paired Normally Distributed Data

For continuous data that has two samples of normally distributed data that is paired, the hypothesis test to use is a paired *t* test.

After the normality check on the differences, as that is the metric of interest with paired data, you can move on to running the paired *t* test. This test requires an alternative hypothesis, a statement that the data is paired, and a confidence level.

To plot the data the R package **PairedData** and its dependencies will be required.

Example 6.8 shows how to run a paired *t* test on data involving performance before and after training.

EXAMPLE 6.8

The customer wants to know if subjects can complete a task quicker after training at the 95% confidence level. Each subject completed a task before training and then completed a similar task after training and the completion time in minutes was recorded. There were 9 subjects and the raw data will be shown in the following example.

```
# Set up the data
Subject = c(1:9)
BeforeTraining = c(15.02, 18.54, 17.66, 16.75, 13.60, 18.30, 14.34,
        18.94, 16.71)
AfterTraining = c(10.83, 16.47, 12.89, 12.46, 13.70, 15.95, 15.56,
        16.32, 13.84)
data16 = data.frame(Subject, BeforeTraining, AfterTraining)

# Load library and check for normality - on the differences
library(car)
qqPlot((data16$BeforeTraining - data16$AfterTraining),
        dist = "norm", main = "Q-Q Plot for Normality",
        xlab = "Norm Quantiles", ylab = "Time Difference (mins)")
```

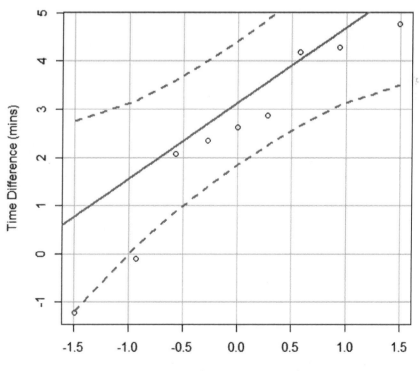

A quantile comparison plot of the differences shows the data to be approximately normally distributed that the line is roughly from the bottom left of the plot to the top right.

```
# Run the paired t test
t.test(data16$BeforeTraining, data16$AfterTraining,
    alternative = "greater", conf.level = 0.95, paired = TRUE)
```

```
Paired t-test

data:  data16$BeforeTraining and data16$AfterTraining
t = 3.6331, df = 8, p-value = 0.003328
alternative hypothesis: true difference in means is greater than 0
95 percent confidence interval:
1.184611       Inf
sample estimates:
mean of the differences
      2.426667
```

```
# Load library and create plot of results
library(PairedData)
paired.plotMcNeil(data16, "BeforeTraining", "AfterTraining",
    subjects = "Subject") + theme_bw() +
    scale_colour_manual(values = c("red", "blue")) +
    scale_x_continuous(limits = c(10,20), breaks = seq(10,20,
        by = 2)) + ylab("Subject") +
    xlab("Time to Complete a Set Task (mins)") +
    ggtitle("Time to Complete a Set Task Before and After Training")
```

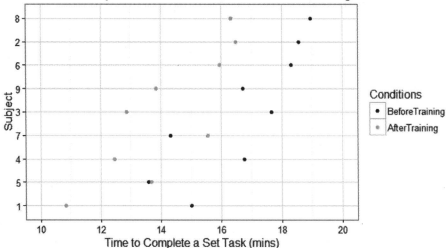

The p-value is 0.003, which is smaller than our significance level of 0.05, which suggests there is evidence to reject the null hypothesis at 95% confidence. In terms of answering the question we can say that there is evidence that training significantly reduces the time taken to complete the task, with an explanation of why "reduces" was stated given in the paragraph following the example.

The output also shows the sample mean of the differences between the before and after completion times (which is 2.43 minutes), which means that subjects are significantly quicker after training on average by 2.43 minutes; it would be a negative value if they were slower. The 95% lower confidence limit of this average difference is 1.18 minutes.

By wanting to know if training made participants complete the task quicker in Example 6.8 this makes the null hypothesis that the training had either no effect or a negative effect, that is, they completed the task in a slower time. Therefore the alternative hypothesis is that they performed the task in a quicker time after training.

If you think about time, if you want before training to be the slower time and after training to be the quicker time, then before training will be a larger number than after training: so before minus after will be a positive number. Therefore the alternative hypothesis is that the difference in means is greater than 0, as that equates to a positive effect from training, which is alternative is **"greater"** in the R code.

As a side note, if you had chosen to do after training minus before training this would make the difference negative. Therefore the alternative hypothesis would need to be that the difference in means is less than 0, which would make the alternative **"less"** in the R code.

The conclusion from the example is that there was evidence that training significantly decreased the time taken to complete a task at the 99% confidence level with a lower bound on the average difference of 1.18 minutes.

Paired Non-Normally Distributed Data

For continuous data that has two samples of non-normally distributed data that is paired, the hypothesis test to use is a Wilcoxon signed-rank test.

Once the normality check on the differences, as this is the metric of interest, has been run and a transformation has been tried, you can move onto running the Wilcoxon signed-rank test. This test requires an alternative hypothesis, a statement that the data is paired, and a confidence level.

The same extra specifications apply as with the previous use with regards to including a pseudomedian of the difference, confidence intervals, and using the exact calculation.

To plot the data the R package **PairedData** and its dependencies will be required.

Example 6.9 looks at running a Wilcoxon signed-rank test using paired data obtained from testing which cage the mice prefer.

EXAMPLE 6.9

The customer wants to know if there is a difference in which cage the mice prefer at the 85% confidence level, the confidence level was set at 85% due to the low cost implications of purchasing new cages. Each mouse spent 30 minutes in each cage and their behavior was monitored through various metrics with the output being combined to give total time spent relaxed and was rounded to the nearest 15 seconds. There were 9 mice involved in the study and the raw data will be shown in the following example.

```
# Set up the data
Mouse = c(1:9)
CageA = c(27.5, 10.25, 24.25, 20.5, 23.75, 25.0, 26.25, 18.25, 10.0)
CageB = c(17.0, 9.25, 16.75, 9.5, 26.75, 23.25, 15.25, 18.0, 12.75)

# Load the library and check for normality - on the differences
library(car)
qqPlot(CageA - CageB, dist = "norm", main = "Q-Q Plot for Normality",
    xlab = "Norm Quantiles", ylab = "Time Difference (mins)")
```

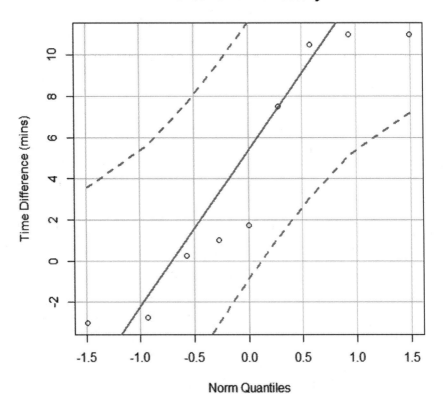

A quantile comparison plot of the differences shows the data is not normally distributed as the line is not straight from the bottom left of the plot to the top right, it is curved.

```
# Assume the transformation didn't work and run the Wilcoxon signed-rank test
wilcox.test(CageA, CageB, alternative = "two.sided", paired = TRUE,
        exact = TRUE, conf.int = TRUE, conf.level = 0.85)
```

```
Wilcoxon signed rank test with continuity correction

data:  CageA and CageB
V = 36, p-value = 0.1232
alternative hypothesis: true location shift is not equal to 0
85 percent confidence interval:
0.2500363   7.4999727
sample estimates:
(pseudo)median
   4.124959
```

```
# Load the library and create plot of results
library(PairedData)
data17 = data.frame(CageA, CageB, Mouse)
paired.plotMcNeil(data17, "CageA", "CageB", subjects = "Mouse") +
        theme_bw() + scale_colour_manual(values = c("red", "blue")) +
        scale_x_continuous(limits = c(0,30), breaks = seq(0,30,
        by = 5)) + ylab("Mouse") +
        xlab("Time Spent Relaxed (mins)") +
        ggtitle("Time Spent Relaxed in Cage A and Cage B")
```

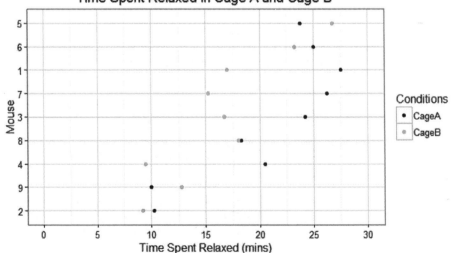

The p-value is 0.123, which is smaller than our significance level of 0.15, which suggests there is evidence to reject the null hypothesis at 85% confidence. In terms of answering the question we can say that there is evidence of a significant difference between the cages with Cage A being the preferred cage.

The output also shows the pseudomedian of the differences between the time spent relaxed within each cage which is 4.12 minutes, which means that the mice are significantly more relaxed in Cage A on "average" by 4.12 minutes. The 85% confidence interval around this difference is 0.25 minutes to 7.50 minutes.

Wanting to know if there was a difference in time spent relaxed in Example 6.9 this makes the null hypothesis that there was no difference between the cages, and the alternative hypothesis that there was a difference between the cages, hence the alternative being *"two.sided"* in the R code.

The conclusion from the example is that there was evidence of a significant difference between the cages at the 85% confidence level, with Cage A being the preferred cage. Cage A should be purchased and the "average" difference in time spent relaxed is between 15 seconds to 7.5 minutes better than Cage B.

Summary

The beginning of the chapter described what hypothesis testing actually is: A way of testing whether a difference was likely due to chance or likely due to a significant difference. It then split into two main sections, the components of hypothesis testing and the methods that can be used to carry out the testing.

The components of hypothesis testing were broken into seven sections; hypotheses, sides and tails, p-values, significant differences, practical differences, plots, and interpretation.

Hypotheses looked at the two competing hypotheses used in testing, the null hypothesis and the alternative hypothesis, with the former being the "take no action" case and the latter being the "take action" case.

Sides and tails was concerned with whether the test is being used to look for a difference, regardless of direction, or that you are interested in the sample being better, or worse, than a value or another sample.

The p-values section detailed the correct definition of what a p-value actually represents in its "backward" language and showed an example to better explain the quite complex value.

Significant differences described how differences are termed as significant, by comparing a p-value to a specific significance level. It also showed that there are multiple ways to quote a significant difference, through confidence levels, significance levels, or strength of evidence.

Practical differences highlighted that it's not all about statistical differences, the practical size of the differences also need to be taken into consideration when forming conclusions.

Plots showed that just because two box plots do not overlap doesn't necessarily mean there is a significant difference between the samples, a hypothesis test still needs to be run.

Interpretation explained the steps to go through to convert the test output into answering the customer question.

The second section detailed the methods that can be used, initially showing a flowchart marking which data types and number of samples require which test. It also clarified that this chapter is only concerned with data that has one or two samples as Chapter 7 deals with bigger data sets and the issue of multiple testing. This was then broken into nine sections corresponding to some possible combinations of data types and samples.

The first three concerned binary data with one sample, two samples, and paired data. They indicated that the three hypothesis test methods to use are the exact binomial test, Fisher's exact test, or chi-square test dependent on cell counts, and the McNemar test, respectively. Each section also contained details of the method and a worked example in R with conclusions.

The next two sections investigated both cases of continuous data of one sample in which the data was normally distributed and where it wasn't and couldn't be transformed. They showed that the hypothesis test methods to use are the one-sample t test and the one-sample Wilcoxon signed-rank test, along with descriptions and worked examples for each.

The following two sections looked at both cases of continuous data with two samples, again normally and non-normally distributed data. They explained that there was a test to run before the hypothesis test, which is a test for equal variance, with Bartlett's being used in the first case and Levene's being used in the second. Then they presented the hypothesis test methods to use for each case, the two-sample t test and the Mann–Whitney test with details and examples of each.

The last two sections described both cases of continuous data with paired samples, with the hypothesis test methods to use being the paired t test and the Wilcoxon signed-rank test. Once again, both sections had descriptions of the method and a worked example.

Chapter 7 progresses on from hypothesis testing on small data sets to hypothesis testing on larger data sets through statistical modeling. The complexity comes from having more variables with probably more levels within each variable, understanding which models to use based on the data collected, and considering the assumptions associated with each. However models are still concerned with hypothesis testing and the output from most of the models will include a p-value, so it is just an extension of what was learned in this chapter.

Statistical Modeling

What Is Actually Going On in the Data?

Statistical modeling is the next step up from simple hypothesis tests; it involves a larger number of explanatory variables, a larger number of levels within each explanatory variable, or both. At this stage we are still only concerned with one response variable.

Models are fit to test the importance of different variables in relation to the response variable, to predict future outcomes, and to assign uncertainty and repeatability to the results.

For example, you may want to test if a variable has a significant effect on an outcome, or you could test what outcome may occur given an untested level of an explanatory variable, or you can assign uncertainty around the model estimates and assess how well the model fitted the data.

There are numerous statistical models and I can't cover all the options, so I have chosen some of the more commonly used ones to discuss. Most of the models will have a similar output; the key is to use the correct model dependent on your data type, led principally by the response variable. In addition all the models will have certain assumptions that need to be satisfied.

©Victoria Cox 2017
V. Cox, *Translating Statistics to Make Decisions*, DOI 10.1007/978-1-4842-2256-0_7

The first section of this chapter looks at the generics of model assumptions, model structure, and model output. The second then delves into some of the commonly used models with explanations of their purpose, R examples, and interpretation of the output.

Statistical Model Components

The key elements involved when using a statistical model for analysis is the input, the methodology, the assumptions, and the output.

Initially when conducting exploratory data analysis (EDA) you should have identified the data types of your response variable and your explanatory variables. It is this information that will determine which methodology to use to create the statistical model—this is discussed more in the Statistical Models section.

Once the correct model has been chosen you will need to confirm that the data satisfies the assumptions of that model type. Depending on sample size and balance of design, some assumptions could be violated without making the model output invalid.

If the assumptions have been satisfied to a certain degree, then the model needs to be formed: the structure. Regardless of the model chosen, this will have a fairly consistent formula to follow.

When the initial model is fitted, it is unlikely that this will be the final model quoted, so you need to follow the process to simplify or expand the model to its best fit.

Finally the model will produce output that will need to be translated correctly for the customer. Again, the model output will be fairly consistent regardless of model type. Most models in this chapter will produce a p-value as with the hypothesis test methods in Chapter 6, along with other useful information.

Model Assumptions

One of the main assumptions for all the models is that the correct data types are being used with the correct models.

The main set of models used are called linear models (LMs), this title contains many different types of model, which is confusing as it is not just your straight line fit. With all types of LMs the following assumptions should be met:

- The relationship between the response variable and the explanatory variable should be approximately linear.

- The residuals should be independent: residuals are the difference in measurements and model estimates.

- The residual variation should be normally distributed.

- The variability in the errors should be consistent: they shouldn't depend on a latent variable or time.

Residuals are the difference between the observed values of the variable and the predicted values from your model.

For the first assumption a simple plot of the data will suffice, Figure 7-1 shows data that satisfies the linearity assumption and data that does not.

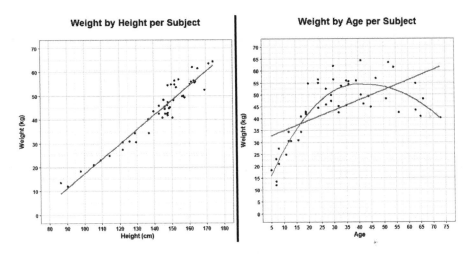

Figure 7-1. Scatter plots to verify linearity assumptions

The residuals being independent include having little or no autocorrelation. This means that a data point is not dependent on the previous data point; for example, stock prices would not have independent residuals due to the current price being linked to the previous price.

The last two assumptions can only be checked after running the model, and the checks are done through drawing objects called diagnostic plots; Figure 7-2 shows one of the plots in a diagnostic plot.

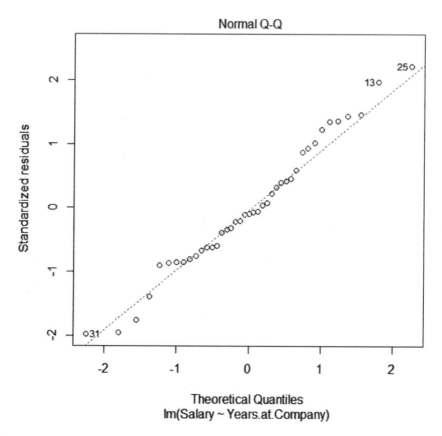

Figure 7-2. An example of a residual normality plot from a diagnostic plot

This plot shows the residuals and should be approximately normally distributed, by showing a straight line from the bottom left to the top right as described in previous chapters.

Figure 7-3 shows the other important plot in a diagnostic plot that concerns the variability in the errors.

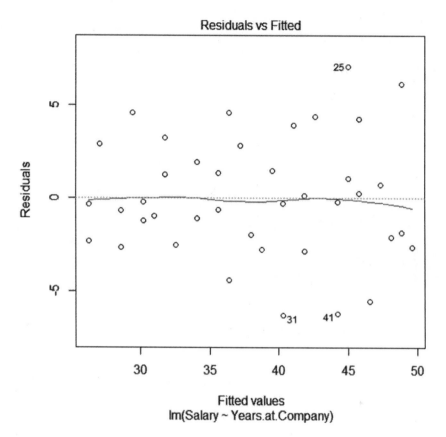

Figure 7-3. An example of an errors variability plot from a diagnostic plot

The points on this plot should be roughly scattered with no discernible pattern, with the red line being fairly horizontal at $y = 0$. If the points are close together on the left and spread out on the right, this suggests that the variation is not consistent, that is, time may be having an effect on the variation. In addition if there are other clear patterns this may be due to another unknown variable. There can be some slight pattern involved vertically, that is stripes, if the explanatory variables are categorical. Figure 7-4 highlights time having an effect on the variation and an acceptable pattern due to categorical variables.

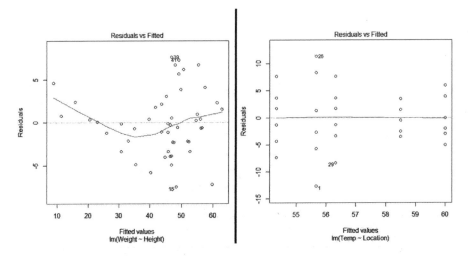

Figure 7-4. Example of an unacceptable and acceptable independence plot

When creating a diagnostics plot, sometimes four plots can be drawn; however these are the two important plots.

Another assumption for all models is that the model is a good fit for the data; again this is another assumption that can only be checked postmodel, there are more in the examples in the Statistical Models section.

Model Structure

The most basic model structure involving a response and an explanatory variable is $y \sim x$. So here you are testing to see if group x has a relationship with the outcome y.

This can get more complicated by adding:

- More explanatory variables: $y \sim x1 + x2$.
- Interactions: $y \sim x1*x2$.
- Polynomials: $y \sim (x1)^2 + x1$.

An interaction, if you recall from Chapter 1 is when the outcome depends on a combination of variables. For example, the outcome may depend on both humidity and kit type: Type A may give a better outcome in humid conditions whereas Type B may give a better outcome in dry conditions.

Polynomials are a confusing case as while they fit a nonlinear shape to the data, the parameters themselves are linear; therefore polynomials are classed as a special type of LM. A first degree polynomial includes main effects and looks like a straight line, a second degree polynomial includes squared terms and is curved, a third degree polynomial includes cubed terms can look like an "S", and so on. Figure 7-5 shows examples of polynomials.

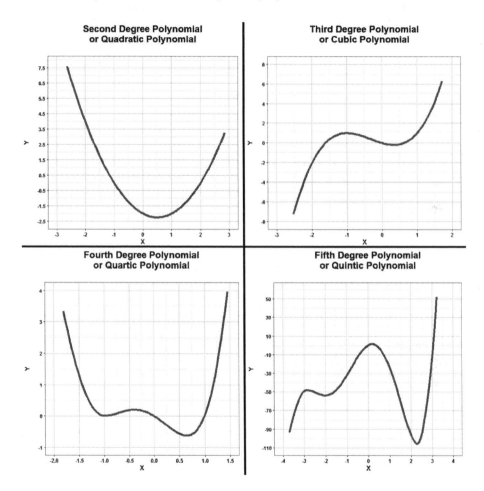

Figure 7-5. Examples of polynomial models

All those three model structures listed earlier—more variables, interactions, and polynomials, can clearly include even more variables and a multitude of levels within each variable.

Model structures can get even more complicated by including mixed effects or nesting. Mixed effects models contain a random effect such as the same subjects repeating the experiment for each group or treatment. This particular example also can be termed repeated measures. Nested models contain variables within other variables, for example class is nested within school.

Model Process

Models can either be built up or down, personally I find it easier to start with the most complex model and simplify it down. However it does depend on the number of variables you begin with.

Let's say we had a model that we knew roughly followed a straight line, so no polynomials, but it could include interactions and our most complicated model was $y \sim x1*x2*x3$. For the example we will look at below, the y and xs translate to "Weight ~ Height*Age*Gender."

This model includes all main effects, all two-way interactions and the three-way interaction: "Height," "Age," "Gender," "Height*Age," "Height*Gender," "Age*Gender," and "Height*Age*Gender." Generally when simplifying a model you would start by excluding the term with the largest p-value with the most complicated interaction.

In this example, if the three-way interaction p-value was significant then we cannot simplify the model as there is evidence that the three-way interaction is having a significant effect on the response, a three-way interaction or higher is quite hard to translate. In some cases it may be simplified, but this is down to judgment about the validity of the three-way interaction and also whether any two-way interactions are actually having a larger effect on the response compared to the three-way interaction.

If the p-value for the three-way interaction didn't imply significance then we could simplify the model. But remember to add in all the two-way interactions again, so our new model would be $y \sim x1*x2 + x1*x3 + x2*x3$.

This model has all the main effects, as all the terms are listed in the model and all the two-way interactions as well. Consider Figure 7-6 for two examples of model output at this stage, only the p-values are shown here.

```
Call: lm(formula = Weight ~ Height*Age + Height*Gender +
    Age*Gender, data = data)
```

Height	0.001 **
Age	0.360
Gender	0.049 *
Height:Age	0.326
Height:Gender	0.037 *
Age:Gender	0.896

Height	0.001 **
Age	0.659
Gender	0.047 *
Height:Age	0.273
Height:Gender	0.041 *
Age:Gender	0.196

Figure 7-6. Example model output

In the output on the left the largest p-value happens to be for a two-way inter-action "Age*Gender," which is in boldface, so that would be the next term to exclude when refitting the model (highlighted in yellow). Remember to only exclude one term at the time, so although there are two, two-way interactions with a nonsignificant p-value: "Age*Gender" and "Height*Age," only the larg-est should be removed. In some cases, terms that were borderline significant, to the level you are working at, can actually become significant as the model is simplified.

In the output on the right the largest p-value is for a main effect "Age," which is in boldface, however this term should not be removed yet as there are still nonsignificant two-way interactions. Remember it is easier to explain a main effect than an interaction. So the term to be removed is the two-way interac-tion with the biggest p-value, which is "Height*Age" (highlighted in yellow).

You may move on to excluding main effects terms once the two-way interac-tions have been cut to only include those with a significant p-value. You can only remove the nonsignificant main effects that are not included in the inter-action. Sometimes you may have a main effect that is not significant, but it is significant in the interaction, this can get tricky to explain.

Once you have simplified your model as far as you can you will need to, check the assumptions, look at the goodness of fit for the model, and also compare the current model to previous models to verify you were justified in simplify-ing it—more in the Statistical Models section.

Further tests may need to be performed such as least-squares means con-trasts. The summary of a model will show that the explanatory variable had an effect on the response. However if there are more than two levels within the explanatory variable, then you don't know which ones are having the effect. This is where the least-squares means contrasts will compare each level of the explanatory variable to each other in terms of the response, it also will correct for conducting multiple testing.

Model Output

Most models will show similar things in the output: information about the residuals, estimates for the variables, errors for the variables, p-values for the variables, information about the degrees of freedom, and information about the goodness of fit.

Figure 7-7 shows some example LM output.

```
Call:
lm(formula = Bottle.Weight ~ Type, data = data)

Residuals:
     Min       1Q    Median       3Q      Max
-1.35789 -0.64789 -0.06842  0.58947  2.18211

Coefficients:
             Estimate  Std. Error  t value  Pr(>|t|)
(Intercept)  21.8105      0.1994   109.368   < 2e-16 ***
TypeB         2.5579      0.2820     9.070  1.93e-12 ***
TypeC         0.3774      0.2820     1.338     0.186
---
Signif. codes:  0 '***' 0.001 '**' 0.01 '*' 0.05 '.' 0.1 ' ' 1

Residual standard error: 0.8693 on 54 degrees of freedom
Multiple R-squared:  0.6397,    Adjusted R-squared:  0.6264
F-statistic: 47.94 on 2 and 54 DF,  p-value: 1.069e-12
```

Figure 7-7. Example linear model output

Here the most important sections of the output have been highlighted. The yellow output shows p-values, these are calculated from testing the effect of the levels of the explanatory variable "Type" *on the response* "Bottle.Weight" in relation to each other. The significant intercept means that the "Bottle. Weight" for "TypeA" is significantly different to zero. R orders factors alphabetically. In this example it has taken out "TypeA" so as not to over fit the model. Here "TypeB" is significantly different to "TypeA" at the 99% confidence level, but "TypeC" is not significantly different to "TypeA". In this scenario you would also want to test for the difference between "TypeB" and "TypeC", and this is done with multiple comparison testing, which can be seen later in Figure 7-8.

The green output shows the estimates, or coefficients, so the intercept is either where the data crosses the y-axis (with a continuous explanatory variable) or the response value for the first level of the factor (with a categorical explanatory variable). The estimates of levels of "Type" show the increase or decrease in "Bottle.Weight":

- For "TypeA" the "Bottle.Weight" will be 21.81g.

- For "TypeB" the "Bottle.Weight" will be (21.81 + 2.56) = 24.37g.

- For "TypeC" the "Bottle.Weight" will be (21.81 + 0.377) = 22.18g; this also highlights why there's not a significant difference between "TypeA" and "TypeC."

The blue output shows how much of the variation has been explained by the model, how well the model has fit the data. In this case 62.6% of the variation has been explained by the model, which is quite good.

In addition you would run the model diagnostics to check the assumptions, calculate confidence intervals for the estimates, determine which "Type" was significantly different to which, and plot the data, which is all shown in Figure 7-8.

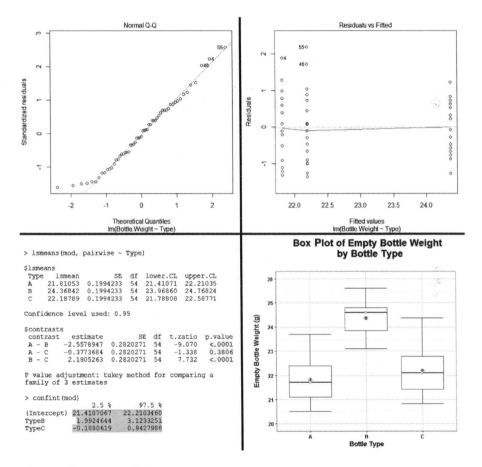

Figure 7-8. Example of additional model output

The top left plot indicates that the model residuals can be assumed approximately normal, and the top right plot shows a random scattering of the points. Remember for categorical variables there won't be any horizontal variation.

The bottom left plot shows the multiple comparisons test using least-squares means contrasts with the Tukey method correcting for multiple comparisons; this is suitable for balanced designs. Another commonly known method for controlling for multiple testing is Bonferroni corrections.

The p-values show that "TypeB" is significantly different to both "TypeA" and "TypeC" at the 99% confidence level, but there is no significant difference between "TypeA" and "TypeC."

Below this are 95% confidence intervals on the estimates, so whilst the average "Bottle.Weight" estimates have been given in the model output, this shows the confidence intervals around those estimates. If you want specific confidence intervals around average values for each level, the best to use are those shown in the top section of the *lsmeans()* output. So although the average values are 21.81g, 24.37g, and 22.18g for "TypeA," "TypeB," and "TypeC," respectively, these three values could be as low as 21.41g, 23.97g, and 21.79g or as high as 22.21g, 24.77g, and 22.59g.

The bottom right plot shows the data visually and helps to explain the direction of the differences between the "Types," it is clear to see why there is no significant difference between "TypeA" and "TypeC." It also clarifies the direction of the difference for "TypeB," it gives significantly heavier "Bottle.Weight" than the other "Types." It also highlights that the mean and median for each "Type" are similar to each other, which is another check for normality.

Statistical Models

The statistical model to use will mainly depend on the data type of the response variable, but can sometimes be dependent on the explanatory variables data type.

There will be other types of models that aren't listed in this chapter as there are so many, however these will be the ones most commonly used.

Each section will discuss the reason to use the model, its assumptions, how to construct the model using an R example, and how to translate the results.

Simple Models

These simple models are the most well-known models, the linear model (LM) and analysis of variance (ANOVA). I have included examples due to this reason; however there really aren't major differences between these two options. In addition, there is also another model that is very similar to the LM called the generalized linear model (GLM).

Linear Model

LMs fit a number of explanatory variables against a continuous response variable. The explanatory variables can be discrete or continuous. In a lot of places you will see recommendations to have more continuous than discrete variables, but this is not necessary.

The assumptions of a LM have been discussed earlier in the chapter, but as a recap they are linearity, independence, residual normality, and consistent error variation.

Example 7.1 will walk through a LM example in R, looking at whether concentration "Conc" has an effect on yield "Yield."

In the interest of space when plotting the graphs in any of the examples in this chapter, only the basic *ggplot()* code will be shown; however the plots I create will include more detail than the basic code produces, this detail will be explained in Chapter 9. In addition to this, usually you would plot basic graphs for EDA at the beginning of the analysis along with well-presented graphs to visually complement the output summary; to save space I will only be creating the final plot. To plot the graphs we will be using the R package *ggplot2*, which have a lot of dependencies.

EXAMPLE 7.1

We have a response variable of Yield and an explanatory variable of Concentration; both the response and explanatory variables are continuous. We are looking to see if Concentration has an effect on Yield.

```
# Input the data
Yield = c(498,480.3,476.4,546,715.4,666,741.2,522,683.6,574,804,637,
          700,750,600,650,590)
Conc = c(3.9,3.8,3.6,4.2,5.7,5,5.5,3.7,4.9,4,6,5,5.2,5.9,4.8,4.7,4.3)
data18 = data.frame(Yield, Conc)

# Fit the model and print the output
mod = lm(Yield ~ Conc, data = data18)
summary(mod)
```

```
Call:
lm(formula = Yield ~ Conc, data = data18)

Residuals:
    Min      1Q  Median      3Q     Max
-35.737 -23.542   5.458  19.435  37.481

Coefficients:
             Estimate  Std. Error  t value  Pr(>|t|)
(Intercept)    40.425      40.338    1.002     0.332
Conc          124.023       8.442   14.692   2.6e-10 ***
---
Signif. codes:  0 '***' 0.001 '**' 0.01 '*' 0.05 '.' 0.1 ' ' 1

Residual standard error: 26.43 on 15 degrees of freedom
Multiple R-squared:  0.935,     Adjusted R-squared:  0.9307
F-statistic: 215.8 on 1 and 15 DF,  p-value: 2.602e-10
```

The model output shows that Concentration is having a significant positive effect on Yield at the 99% confidence level, p-value of 2.6e-10. For every 1 unit increase in Concentration there is a 124 unit increase in Yield the estimate by Conc is 124.023. It also shows that the model accounts for 93.1% of the total variation, which is very good—adjusted R^2 of 0.9307.

```
# Check the diagnostics
plot(mod, which = 2)
plot(mod, which = 1)
```

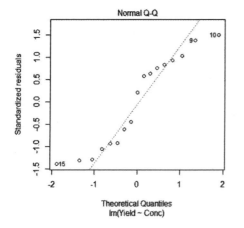

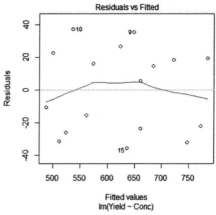

The diagnostic plots show that we can roughly assume normality on the residuals, it's not great but it is good enough, and that there is consistent error variation as the points are roughly scattered.

```
# Calculate confidence intervals for the estimates
confint(mod)

                2.5 %      97.5 %
(Intercept)   -45.55406   126.4039
Conc          106.03008   142.0167
```

The confidence intervals highlight that for every unit increase in Concentration the increase in Yield could be between 106 and 142. The confidence intervals for the intercept also include 0, which is good as we would expect a Yield of 0 at a Concentration of 0.

```
# Plot the data
library(ggplot2)
ggplot(data18, aes(x = Conc, y = Yield)) + theme_bw() +
        geom_point() + geom_smooth(method = "lm")
```

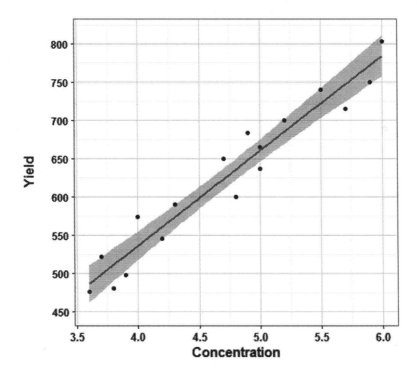

Yield by Concentration

Finally the scatter plot highlights the linearity of the data and shows the line of best fit with confidence intervals. This clearly emphasises the positive relationship between Concentration and Yield.

Once the data has been input, you assign a model to the data and call it a name, in this case "mod." The **lm( )** stands for linear model, then the response variable is specified followed by the explanatory variable(s), finally the data set to be used is included.

The next few commands are carried out on this "mod" object, so the **summary( )** gives a summary of the model output. Next the diagnostic plots are drawn, by specifying the "**which =**" part allows us to define which diagnostic plot we would like drawn. The **confint( )** calculates 95% confidence intervals by default of the model estimates, to change this level you would amend the code as follows: **confint(mod, level = 0.99)**.

Finally using **ggplot( )** I plotted a scatter plot including the line of best fit from the linear model that has 95% confidence interval shading.

As an aside, a model can be fitted without an intercept term. This just means that the slope of the fit is forced through the origin, (0,0), which in some cases makes sense but in others may not.

ANOVA

Many people have heard of ANOVA, it is simply an extension of a two-sample *t* test. As with LMs, they also are used to fit a number of explanatory variables against a continuous response variable. Again, the explanatory variables can be discrete or continuous. However in a lot of places you will see recommendations to have more discrete variables than continuous variables, this should be adhered to, otherwise you can just run a LM instead.

The key thing to note here is that there isn't much difference between fitting a LM or an ANOVA. You can in fact get the same ANOVA results by computing an ANOVA table of the LM results. The main difference between the two is that with a LM it doesn't matter what order you put the explanatory variables into the model, whereas it does matter for the general ANOVA model.

The LM will use t-values to calculate the p-values; the t-values test the marginal impact of the levels of the explanatory variables given the fact that all the other variables are present. The ANOVA will use F-values to calculate the p-values; the F-values test whether the explanatory variable as a whole reduces the residual sum of squares (SS) compared to the previous explanatory variable(s). So for explanatory variable one this will be tested against the response, then explanatory variable two will be tested given explanatory variable one is present, and so forth.

An advantage of the ANOVA table is that it can tidy up high level explanatory variables. A LM will show the estimate parameters, that is, one row per explanatory variable level, whereas an ANOVA will show the variables, that is, one row per explanatory variable. For example, a LM output of one explanatory variable with 10 levels will show 9 rows, whereas an ANOVA will only show 1 row. It can be simpler to use an ANOVA table to simplify the model; however I would compare the two along the way, and always use a LM output for the final results.

The assumptions of the ANOVA are similar to that of the LM including independence, consistent error variation, and residual normality. It also includes the assumption that the levels of the explanatory variable have similar variation. The recommended test to use for this is Bartlett's, along with plotting a box plot, recall Chapter 6.

Example 7.2 will look at an ANOVA example in R, looking at whether the either of the two materials "Material" or any of the four methods of implementation "Method" have an effect on the total volume produced "Volume." It will also show a LM output of the same model along with how you can calculate an ANOVA table from that output.

As there are multiple "Methods" we may need to carry out multiple comparisons, if so we will need to use the R package *lsmeans*. Least-squares means can be calculated for each explanatory variable combination along with their contrasts to determine if there are significant differences between multiple groups.

EXAMPLE 7.2

We have a response variable of Volume and two explanatory variables of Material and Method. The response variable is continuous and both explanatory variables are discrete. We are looking to see if Material, Method, or an interaction between the two has an effect on Volume.

```
# Input the data
Volume = c(28.756,29.305,28.622,30.195,27.736,17.093,17.076,17.354,
           16.353,15.880, 36.833,35.653,34.583,35.504,35.236,30.333,
           30.030,28.339,28.748,29.020,32.591,30.572,32.904,31.942,
           33.653, 20.725,22.198,21.988,22.403,21.324,38.840,40.137,
           39.295,39.006,40.731,32.136,33.209,34.558,32.782,31.460)
Material = rep(c("A","B"), each = 20)
Method = rep(c("I","II","III","IV"), each = 5, 2)
data19 = data.frame(Volume, Material, Method)

# Check for equal variances
bartlett.test(Volume ~ interaction(Material,Method), data = data19)

Bartlett test of homogeneity of variances

data:  Volume by interaction(Material, Method)
Bartlett's K-squared = 2.6181, df = 7, p-value = 0.9179
```

The Bartlett's test gave a p-values of 0.918 that suggested no evidence to reject the null hypothesis, which means we can assume equal variances and that was also confirmed by the roughly equal sizes of the box and whiskers on the box plot at the end of Example 7.2.

```
# Fit the full model and print the output
mod = aov(Volume ~ Material*Method, data = data19)
summary(mod)
```

	Df	Sum Sq	Mean Sq	F value	Pr(>F)	
Material	1	159.2	159.2	197.662	3.03e-15	***
Method	3	1742.4	580.8	721.025	< 2e-16	***
Material:Method	3	3.8	1.3	1.572	0.215	
Residuals	32	25.8	0.8			

```
---
Signif. codes:  0 '***' 0.001 '**' 0.01 '*' 0.05 '.' 0.1 ' ' 1
```

The output of the full ANOVA model that contains the interaction showed that the interaction was not significant to Volume with a p-value of 0.215. This is backed by the plot of the data that shows a similar pattern and gradient for Material and Method. As such the model can be simplified by removing the interaction term.

```
# Simplify the model and print the output
mod2 = aov(Volume ~ Material + Method, data = data19)
summary(mod2)
```

	Df	Sum Sq	Mean Sq	F value	Pr(>F)	
Material	1	159.2	159.2	188.4	1.18e-15	***
Method	3	1742.4	580.8	687.3	< 2e-16	***
Residuals	35	29.6	0.8			

```
---
Signif. codes:  0 '***' 0.001 '**' 0.01 '*' 0.05 '.' 0.1 ' ' 1
```

Once simplified the model output showed that both Material and Method had a significant effect on Volume at the 99% confidence level, again it's clear to see a difference between the two Methods and you also can see that there is a consistent difference between the Materials using the final plot. This means that the model cannot be simplified any further.

The ANOVA output showed us that there is a difference between Method A and B and looking at the box plot we can see that Method B gives a higher Volume. However we don't know if all Methods are different to each other, and this is what the lsmeans output will show us.

```
# Check for differences between all Methods
library(lsmeans)
lsmeans(mod2, pairwise ~ Method)
```

```
$lsmeans
```

Method	lsmean	SE	df	lower.CL	upper.CL
I	30.6276	0.2906892	35	30.03747	31.21773
II	19.2394	0.2906892	35	18.64927	19.82953
III	37.5818	0.2906892	35	36.99167	38.17193
IV	31.0615	0.2906892	35	30.47137	31.65163

```
Results are averaged over the levels of: Material
Confidence level used: 0.95

$contrasts
contrast  estimate        SE  df  t.ratio  p.value
 I - II    11.3882  0.4110966  35   27.702   <.0001
 I - III   -6.9542  0.4110966  35  -16.916   <.0001
 I - IV    -0.4339  0.4110966  35   -1.055   0.7183
II - III  -18.3424  0.4110966  35  -44.618   <.0001
II - IV   -11.8221  0.4110966  35  -28.757   <.0001
III - IV    6.5203  0.4110966  35   15.861   <.0001

Results are averaged over the levels of: Material
P-value adjustment: tukey method for comparing a family of 4
estimates
```

Using the second section of the lsmeans output we can see that all Methods are significantly different to each other at the 99% confidence level except Method I and Method IV, which are not significantly different to each other—p-value of 0.7183. Using the first section of the lsmeans output and/or the box plot shown at the end of the example, you also can order the Methods accordingly, from highest Volume to lowest: Method III, Method IV and Method I, then Method II.

```
# Compare the two models
anova(mod, mod2)
```

```
Analysis of Variance Table

Model 1: Volume ~ Material * Method
Model 2: Volume ~ Material + Method
  Res.Df     RSS  Df  Sum of Sq       F  Pr(>F)
1     32  25.777
2     35  29.575  -3    -3.7984  1.5718  0.2154
```

To check the model could be simplified an ANOVA comparing the two models was run, comparing the simpler model to the more complex model. This showed a p-value of 0.215 that suggests no significant difference between the two models, hence it is fine to use the simpler model.

```
# Check the diagnostics
plot(mod2, which = 2)
plot(mod2, which = 1)
```

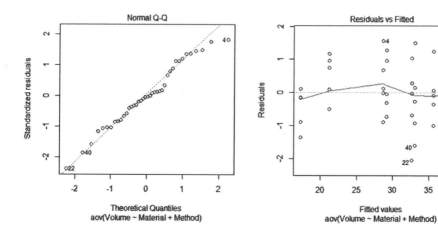

aov(Volume ~ Material + Method)

The diagnostic plots show that we can roughly assume normality on the residuals and that there is consistent error variation, the points are roughly scattered. These plots will be exactly the same if they were run on the LM output.

```
# Calculate confidence intervals for the estimates
confint(mod2)
```

```
                 2.5 %        97.5 %
(Intercept)    27.9726891    29.29226
MaterialB       3.4001196     4.58038
MethodII      -12.2227704   -10.55363
MethodIII       6.1196296     7.78877
MethodIV       -0.4006704     1.26847
```

The downside of the ANOVA summary output is that it doesn't show the estimates for each level of the explanatory variables; however this could be calculated from the confidence intervals output, as they are symmetrical. For example, the estimate for Material A would be $(27.97 + ((29.29 - 27.97)/2)) = 28.63$, which you will see matches the LM output next.

```
# Fit linear model and print results
mod3 = lm(Volume ~ Material + Method, data = data19)
summary(mod3)
```

```
Call:
lm(formula = Volume ~ Material + Method, data = data19)

Residuals:
    Min        1Q     Median       3Q        Max
-2.05073  -0.59837   -0.03905   0.69276   1.56253

Coefficients:
              Estimate  Std. Error  t value  Pr(>|t|)
(Intercept)    28.6325      0.3250    88.100   < 2e-16 ***
MaterialB       3.9903      0.2907    13.727  1.18e-15 ***
MethodII      -11.3882      0.4111   -27.702   < 2e-16 ***
```

```
MethodIII     6.9542      0.4111     16.916     < 2e-16 ***
MethodIV      0.4339      0.4111      1.055     0.298
---
Signif. codes:  0 '***' 0.001 '**' 0.01 '*' 0.05 '.' 0.1 ' ' 1

Residual standard error: 0.9192 on 35 degrees of freedom
Multiple R-squared:  0.9847,    Adjusted R-squared:  0.9829
F-statistic: 562.6 on 4 and 35 DF,  p-value: < 2.2e-16
```

By fitting the LM we can see that the Volume for Material A and Method I is 28.63, the Volume for Material A and Method II is $(28.63 - 11.39) = 17.24$, through to the Volume for Material B and Method IV, which is $(28.63 + 3.99 + 0.43) = 33.05$. We also can see that the model explains 98.3% of the variation.

```
# Calculate confidence intervals - only first section of output shown
lsmeans(mod2, pairwise ~ Method*Material)

$lsmeans
Method Material    lsmean        SE df lower.CL upper.CL
   I         A   28.63247 0.3250004 35 27.97269 29.29226
  II         A   17.24427 0.3250004 35 16.58449 17.90406
 III         A   35.58667 0.3250004 35 34.92689 36.24646
  IV         A   29.06637 0.3250004 35 28.40659 29.72616
   I         B   32.62272 0.3250004 35 31.96294 33.28251
  II         B   21.23452 0.3250004 35 20.57474 21.89431
 III         B   39.57692 0.3250004 35 38.91714 40.23671
  IV         B   33.05662 0.3250004 35 32.39684 33.71641
```

If we wanted confidence intervals on each group level, we could use the output from a least-squares means output, which includes the interaction term. So the confidence interval values we would get for the three examples above are 27.97 to 29.29, 16.58 to 17.90, and 32.40 to 33.72, respectively.

```
# Create ANOVA table from linear model results
anova(mod3)

Analysis of Variance Table

Response: Volume
          Df  Sum Sq Mean Sq F value    Pr(>F)
Material   1  159.22  159.22  188.43 1.183e-15 ***
Method     3 1742.40  580.80  687.34 < 2.2e-16 ***
Residuals 35   29.58    0.85
---
Signif. codes:  0 '***' 0.001 '**' 0.01 '*' 0.05 '.' 0.1 ' ' 1
```

By creating an ANOVA table from the LM output we can see that it gives exactly the same results as running the original ANOVA model on the data.

```
# Plot data
ggplot(data19, aes(x = Method, y = Volume)) + theme_bw() +
        facet_wrap( ~ Material) + stat_boxplot(geom = "errorbar") +
        geom_boxplot()
```

Volume by Material and Method

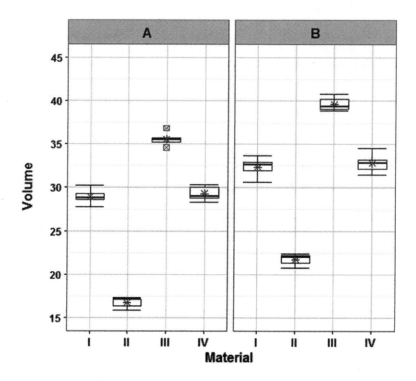

Finally the box plot highlights the differences found between the explanatory variables in terms of Volume: All four boxes in the B box give a higher Volume than their corresponding boxes in the A box, and the order of the Materials is the same regardless of the Method. It also emphasises the similar variation as mentioned previously, in fact showing that there was very little variation due to the boxes and whiskers being so small.

Once the data has been input, you check for equal variance using the Bartlett's test, **bartlett.test( )**, then assign a model to the data and call it a name, which in this case "mod." The **aov( )** stands for ANOVA, then the response variable is specified followed by the explanatory variables and interaction, finally the data set to be used is included.

The next few commands are carried out on this "mod" object or "mod2" object, which was our simplified model so the **summary( )** gives a summary of the model output. The summary can show us differences between the "Materials" as there were only two, but it cannot show us differences between the "Methods." Therefore **lsmeans( )** does that by specifying the model along with which explanatory variable is of interest, in this case "Method."

Next the ANOVA run is to test between the two models, the full model and the simplified model. The diagnostic plots are drawn and by specifying the "*which* =" part in the *plot( )* command allows us to define which diagnostic plot we would like drawn. The *confint( )* calculates 95% confidence intervals by default of the model estimates, to change this level you would amend the code as follows: *confint(mod, level = 0.99)*. For confidence intervals around the average per variable level, the output from the *lsmeans( )* was used.

The LM was fit in the same way as described in Example 7.1 and an ANOVA table of the results was calculated to show their similarities.

Finally using *ggplot( )* I plotted a box plot with the mean and statistical outliers clearly highlighted. As the interaction was not significant I didn't need to plot both variables on one graph, I could have plotted a box plot showing "Material" against "Volume" and another showing "Method" *against* "Volume."

This example highlights that fitting an LM and an ANOVA are very similar procedures. The recommendation would be to use an LM over an ANOVA, except if a long complicated table needs to be tidied up for comprehension.

Generalized Linear Model

GLMs are one step up from a simple LM due to the fact that you can add in a distribution. A GLM will be able to fit a multitude of distributions, some of which include a Gaussian model, a Poisson model, a negative-binomial model, and a binomial model. This lets you model all types of discrete data. However in all cases the set-up of the response and explanatory variables is the same, algebraically and using software.

The model summary output of the default GLM, which is the Gaussian distribution, will be exactly the same as the LM output. The difference will be seen in the confidence intervals of the estimates, this is because the GLM uses the normal distribution whereas the LM uses the t distribution.

When comparing GLM models this is done using the same test as in the ANOVA example, however the test type chosen depends on the data type. For Gaussian GLMs the F test is recommended, whereas for Poisson, negative binomial and binomial GLMs the chi-square test is more appropriate.

Gaussian GLM

The Gaussian GLM is the default if no other distribution is chosen. It will produce the same results as running a simple LM, so it's down to personal choice whether you use a GLM Gaussian distribution or a simple LM. As such, it also has the same assumptions as those of the LM.

Example 7.3 looks at a GLM with a Gaussian distribution example in R, it also shows that running a simple LM gives you the same results. The example investigates whether any of the three machines "Machine," either of the operators "Operator," or the concentration that is recorded as low, medium, or high "Concentration" has an effect on the time taken to find the substance of interest in minutes "Time.Taken."

EXAMPLE 7.3

We have a response variable of Time.Taken and three explanatory variables of Machine, Operator, and Concentration. The response variable is continuous and all three explanatory variables are discrete. We are looking to see if Machine, Operator, Concentration, or any interactions between them has an effect on Time.Taken.

```
# Input the data
Time.Taken = c(48.1,46.3,47.2,47.9,47.6,49,48,38.6,39.8,40.9,41.7,39.9,
               40.8,39.7,33.3,34.6,35.8,34,32,35.4,34.5,44.3,45.8,46.5,48.7,
               45.3,48.8,49.4,38.8,38.2,38.7,39.2,39.1,40,41.9,34.2,33.6,
               33.7,34,34.3,35.1,35.8,33.4,32.1,34.8,35.7,34.6,35.7,38.4,
               30.9,31,30.7,31.1,31.8,32.6,33.2,28,28.6,27.7,28.3,29,28.4,
               29.1,35.8,33.2,37.1,36.4,37.6,38.9,39.2,31.1,30.5,30.4,31.6,
               31.9,32.6,33.7,28.7,28.9,27.8,29.1,29,28.4,28.7,47.6,46.3,
               47.9,46.7,49.8,48.9,49.4,39.1,38.5,38.4,38.7,39.5,40.5,42.3,
               35.3,34.4,34.2,34.2,34.7,35,34.6,47.8,45.1,48.8,48.4,49.7,
               49.2,47,40.1,39.9,39.5,38.9,40.9,40.6,42,34.9,34.2,34.7,35,
               33.8,33.1,34.5)
Machine = rep(c("A","B","C"), each = 42)
Operator = rep(c("Op1","Op2"), each = 21, 3)
Concentration = rep(c("Low","Medium","High"), each = 7, 6)
data20 = data.frame(Time.Taken, Machine, Operator, Concentration)
```

```
# Order the levels of the Concentration variable
data20$Concentration = factor(data20$Concentration,
        levels = c("Low","Medium","High"))
```

R will order the levels of a variable alphabetically; this would mean Concentration would be High, Low, then Medium, which clearly isn't what we want. By using the command above we have reordered the Concentration explanatory variable to the correct order.

```
# Fit the full GLM including all interactions (label abbr. in some cases)

mod = glm(Time.Taken ~ Machine*Operator*Concentration,
        family = gaussian, data = data20)
summary(mod)
```

```
Call:
glm(formula = Time.Taken ~ Machine * Operator * Concentration,
        family = gaussian, data = data20)

Deviance Residuals:
    Min      1Q   Median      3Q     Max
-3.6857  -0.6143  -0.1000  0.6679  3.4429

Coefficients:
                                    Estimate Std. Error  t value Pr(>|t|)
(Intercept)                           47.729      0.473  100.919  < 2e-16 ***
MachineB                             -12.771      0.669  -19.095  < 2e-16 ***
MachineC                               0.357      0.669    0.534   0.5945
OperatorOp2                           -0.757      0.669   -1.132   0.2601
ConcentrationMedium                   -7.529      0.669  -11.256  < 2e-16 ***
ConcentrationHigh                    -13.500      0.669  -20.184  < 2e-16 ***
MachineB:OperatorOp2                   2.686      0.946    2.839   0.0054 **
MachineC:OperatorOp2                   0.671      0.946    0.710   0.4793
MachineB:ConcentrationMedium           4.186      0.946    4.425 2.31e-05 ***
MachineC:ConcentrationMedium          -0.986      0.946   -1.042   0.2997
MachineB:ConcentrationHigh             6.986      0.946    7.385 3.36e-11 ***
MachineC:ConcentrationHigh             0.043      0.946    0.045    0.964
OperatorOp2:ConcentrationMed          -0.029      0.946   -0.030    0.976
OperatorOp2:ConcentrationHigh          0.914      0.946    0.967    0.336
MachineB:OperatorOp2:ConcMed          -1.829      1.338   -1.367    0.175
MachineC:OperatorOp2:ConcMed           0.814      1.338    0.609    0.544
MachineB:OperatorOp2:ConcHigh         -2.629      1.338   -1.965    0.052 .
MachineC:OperatorOp2:ConcHigh         -1.143      1.338   -0.854    0.395
---
Signif. codes:  0 '***' 0.001 '**' 0.01 '*' 0.05 '.' 0.1 ' ' 1

(Dispersion parameter for gaussian family taken to be 1.565714)

Null deviance: 5156.4  on 125  degrees of freedom
Residual deviance:  169.1  on 108  degrees of freedom
AIC: 432.64

Number of Fisher Scoring iterations: 2
```

The output of the full GLM is quite cumbersome to go through at this stage, with this many levels of explanatory variables it's quite difficult to see which terms are significant to the model so we calculate an ANOVA table of the results.

```
# Tidy the output up with an ANOVA table (label abbr. in some cases)
anova(lm(mod))
```

Analysis of Variance Table

Response: Time.Taken

	Df	Sum Sq	Mean Sq	F value	Pr(>F)	
Machine	2	2077.67	1038.84	663.4899	< 2.2e-16	***
Operator	1	0.50	0.50	0.3164	0.57497	
Concentration	2	2722.01	1361.01	869.2552	< 2.2e-16	***

```
Machine:Operator           2     7.57    3.79    2.4175    0.09396 .
Machine:Concentration      4   168.05   42.01   26.8320  1.834e-15 ***
Operator:Concentration     2     0.88    0.44    0.2823    0.75461
Machine:Operator:Conc      4    10.59    2.65    1.6904    0.15747
Residuals                108   169.10    1.57
---
Signif. codes:  0 '***' 0.001 '**' 0.01 '*' 0.05 '.' 0.1 ' ' 1
```

The output of the ANOVA table from the full GLM that contained all the main effects
and interactions showed that the first term to remove is the three-way interaction
between Machine, Operator, and Concentration, p-value of 0.157.

```
# Simplify the model by removing one interaction or explanatory
variable at a time
# Output omitted for space saving - except for the final model
mod2 = glm(Time.Taken ~ Machine*Operator + Machine*Concentration +
        Operator*Concentration, family = gaussian, data = data20)
summary(mod2)
anova(lm(mod2))
```

The next term to be removed once the model was refitted, including the three two-
way interactions, was the interaction between Operator and Concentration, p-value
of 0.760.

```
mod3 = glm(Time.Taken ~ Machine*Operator + Machine*Concentration,
        family = gaussian, data = data20)
summary(mod3)
anova(lm(mod3))
```

Refitting, the next model showed that the interaction between Machine and Operator
could be removed, p-value of 0.096.

```
mod4 = glm(Time.Taken ~ Machine*Concentration + Operator,
        family = gaussian, data = data20)
summary(mod4)
anova(lm(mod4))
```

The next model fit showed that the interaction between Machine and Concentration
was still significant, however the main effect of Operator was not significant, p-values
of 2.373e-15 and 0.582, respectively. As Operator was not included in the interaction
term this could be removed and the model simplified once more.

```
mod5 = glm(Time.Taken ~ Machine*Concentration, family = gaussian,
        data = data20)
summary(mod5)
```

```
Call:
glm(formula = Time.Taken ~ Machine * Concentration, family =
gaussian,
        data = data20)
```

```
Deviance Residuals:
    Min      1Q   Median      3Q      Max
-3.8214  -0.7393  -0.0143  0.7321   3.2786
```

Coefficients:

	Estimate	Std. Error	t value	Pr(>|t|)	
(Intercept)	47.3500	0.3394	139.530	< 2e-16	***
MachineB	-11.4286	0.4799	-23.814	< 2e-16	***
MachineC	0.6929	0.4799	1.444	0.151	
ConcentrationMedium	-7.5429	0.4799	-15.717	< 2e-16	***
ConcentrationHigh	-13.0429	0.4799	-27.177	< 2e-16	***
MachineB:ConcMedium	3.2714	0.6787	4.820	4.35e-06	***
MachineC:ConcMedium	-0.5786	0.6787	-0.852	0.396	
MachineB:ConcHigh	5.6714	0.6787	8.356	1.53e-13	***
MachineC:ConcHigh	-0.5286	0.6787	-0.779	0.438	

```
---
Signif. codes:  0 '***' 0.001 '**' 0.01 '*' 0.05 '.' 0.1 ' ' 1
```

(Dispersion parameter for gaussian family taken to be 1.612253)

```
Null deviance: 5679.24  on 125  degrees of freedom
Residual deviance:  544.71  on 117  degrees of freedom
AIC: 562.03
```

Number of Fisher Scoring iterations: 2

anova(lm(mod5))
Analysis of Variance Table

Response: Time.Taken

	Df	Sum Sq	Mean Sq	F value	Pr(>F)	
Machine	2	2077.67	1038.84	644.338	< 2.2e-16	***
Concentration	2	2722.01	1361.01	844.164	< 2.2e-16	***
Machine:Concentration	4	168.05	42.01	26.058	1.868e-15	***
Residuals	117	188.63	1.61			

```
---
Signif. codes:  0 '***' 0.001 '**' 0.01 '*' 0.05 '.' 0.1 ' ' 1
```

The final model output showed that Machine, Concentration, and the interaction between Machine and Concentration were significant to the model at the 99% confidence level. The residual deviance (544.71) is much smaller than the null deviance (5679.24) that suggests this model is a lot better than the null model.

By looking at the estimates of the GLM we can see that the Time.Taken for Machine-A and Concentration-Low is 47.35, the Time.Taken for Machine-A and Concentration-Medium is (47.35 − 7.54) = 39.81, through to the Time.Taken for Machine-C and Concentration-High, which is (47.35 + 0.69 − 13.04 − 0.53) = 34.47. A table with all the estimates is shown below, this is just to highlight the values you would get by using the estimates output—it won't be repeated in other examples to save space.

Model	Concentration		
Estimates	Low	Medium	High
Machine **A**	47.35	39.81	34.31
Machine **B**	35.92	31.65	28.55
Machine **C**	48.04	39.92	34.47

```
# Calculate estimate confidence intervals
confint(mod5)
```

	2.5 %	97.5 %
(Intercept)	46.6848790	48.0151210
MachineB	-12.3691945	-10.4879483
MachineC	-0.2477659	1.6334802
ConcentrationMedium	-8.4834802	-6.6022341
ConcentrationHigh	-13.9834802	-12.1022341
MachineB:ConcentrationMedium	1.9411866	4.6016705
MachineC:ConcentrationMedium	-1.9088134	0.7516705
MachineB:ConcentrationHigh	4.3411866	7.0016705
MachineC:ConcentrationHigh	-1.8588134	0.8016705

If we wanted confidence intervals on the estimates we can use the earlier values.

```
# Calculate confidence intervals - only first section of output shown
lsmeans(mod5, pairwise ~ Machine*Concentration)
```

```
$lsmeans
Machine    Conc    lsmean          SE df  asymp.LCL  asymp.UCL
      A     Low  47.35000  0.3393537 NA   46.68488   48.01512
      B     Low  35.92143  0.3393537 NA   35.25631   36.58655
      C     Low  48.04286  0.3393537 NA   47.37774   48.70798
      A  Medium  39.80714  0.3393537 NA   39.14202   40.47226
      B  Medium  31.65000  0.3393537 NA   30.98488   32.31512
      C  Medium  39.92143  0.3393537 NA   39.25631   40.58655
      A    High  34.30714  0.3393537 NA   33.64202   34.97226
      B    High  28.55000  0.3393537 NA   27.88488   29.21512
      C    High  34.47143  0.3393537 NA   33.80631   35.13655
```

Results are given on the identity (not the response) scale.
Confidence level used: 0.95

If we wanted confidence intervals on each group level, we could use the output from a least-squares means output that includes the interaction term. So the confidence interval values we would get for the three examples above are 46.68 to 48.02, 39.14 to 40.47, and 33.81 to 35.14, respectively.

```
# Check for differences between all Machines, Concentrations and the
# interaction between the two
# Only contrasts shown, and last lsmeans omitted for space saving
lsmeans(mod5, pairwise ~ Machine)
lsmeans(mod5, pairwise ~ Concentration)
lsmeans(mod5, pairwise ~ Machine*Concentration)
```

```
NOTE: Results may be misleading due to involvement in interactions
$contrasts
contrast    estimate       SE  df  z.ratio  p.value
  A - B    8.4476190  0.2770811  NA   30.488   <.0001
  A - C   -0.3238095  0.2770811  NA   -1.169   0.4720
  B - C   -8.7714286  0.2770811  NA  -31.657   <.0001
```

Results are averaged over the levels of: Concentration
P-value adjustment: tukey method for comparing a family of 3
estimates

```
$contrasts
      contrast    estimate       SE  df  z.ratio  p.value
  Low - Medium    6.645238  0.2770811  NA   23.983   <.0001
    Low - High   11.328571  0.2770811  NA   40.885   <.0001
 Medium - High    4.683333  0.2770811  NA   16.902   <.0001
```

Results are averaged over the levels of: Machine
P-value adjustment: tukey method for comparing a family of 3
estimates

The model output can tell us whether the explanatory variables were significant to the model. However as both Machine and Concentration have more than two levels, lsmeans is required to define where those differences are.

Looking at the lsmeans output for Machine, it showed that Machine-B gave significantly quicker Time.Taken than both Machine-A and Machine-C at the 99% confidence level. There was no significant difference between Machine-A and Machine-C, p-value of 0.472.

The lsmeans output for Concentration showed that all Concentrations were significantly different to each other at the 99% confidence level, with Concentration-High giving the quickest Time.Taken, then Concentration-Medium, then Concentration-Low. However, both these and the Machine values are calculated by averaging over the other explanatory variable due to the interaction term, so the interaction term output is the most important to look through.

The output was not shown due to saving space, but using the p-values and the box plot drawn in the next command as a visual aid, the output is as follows (with M = Machine and C = Concentration):

- "MB–CHigh" gave significantly quicker Time.Taken than all other combinations at the 99% confidence level.

- "MB–CMedium" gave significantly quicker Time.Taken than all other remaining combinations at the 99% confidence level.

- Both "MA–CHigh" and "MC–CHigh" gave significantly quicker Time.Taken than "MA–CLow," "MC–CLow," "MA–CMedium," and "MC–CMedium" at the 99% confidence level. They also gave significantly quicker Time.Taken than "MB–CLow" at the 95% and 90% confidence level, respectively.

- "MB–CLow" gave significantly quicker Time.Taken than "MA–CMedium," "MC–CMedium," "MA–CLow," and MC–CLow" at the 99% confidence level.

- Both "MA–CMedium" and "MC–CMedium" gave significantly quicker Time.Taken than both "MA–CLow" and "MC–CLow" at the 99% confidence level.

```
# Plot the box plot to aid with lmeans interpretation
ggplot(data20, aes(x = Concentration, y = Time.Taken)) + theme_bw() +
    facet_wrap(~ Machine) + stat_boxplot(geom = "errorbar") +
    geom_boxplot()
```

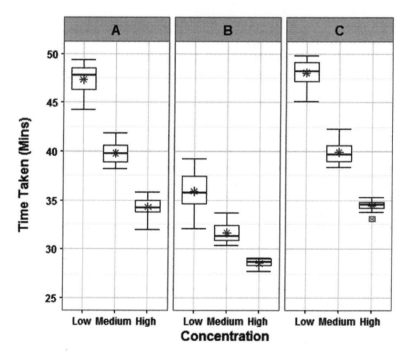

```
# Compare the models
anova(mod, mod2, mod3, mod4, mod5, test = "F")
```

Analysis of Deviance Table

Model 1: Time.Taken ~ Machine * Operator * Concentration
Model 2: Time.Taken ~ Machine * Operator + Machine * Concentration +
        Operator * Concentration
Model 3: Time.Taken ~ Machine * Operator + Machine * Concentration
Model 4: Time.Taken ~ Machine * Concentration + Operator

```
Model 5: Time.Taken ~ Machine * Concentration
  Resid. Df  Resid. Dev  Df  Deviance      F   Pr(>F)
1       108      169.10
2       112      179.68  -4  -10.5870  1.6904  0.15747
3       114      180.57  -2   -0.8840  0.2823  0.75461
4       116      188.14  -2   -7.5702  2.4175  0.09396 .
5       117      188.63  -1   -0.4953  0.3164  0.57497
---
Signif. codes:  0 '***' 0.001 '**' 0.01 '*' 0.05 '.' 0.1 ' ' 1
```

To check the model could be simplified an ANOVA comparing all the models was run
that compared the simpler models to the more complex models. All these p-values
suggest no significant difference between the models, hence it is fine to use the
simpler model; these p-values are actually the p-values of the removed terms.

```
# Check the diagnostics
plot(mod5, which = 2)
plot(mod5, which = 1)
```

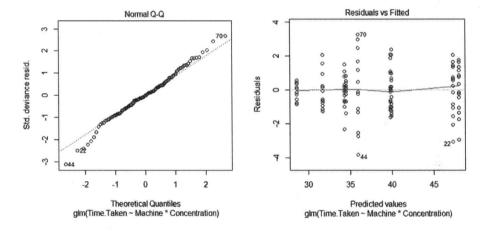

The diagnostic plots show that we can assume normality on the residuals and that
there is consistent error variation as the points are roughly scattered. These plots will
be exactly the same if they were run on the LM output.

```
# Run the simple linear model - output omitted except the last few lines
# due to replication and space saving
summary(lm(mod5))
```

```
Residual standard error: 1.27 on 117 degrees of freedom
Multiple R-squared: 0.9634,    Adjusted R-squared: 0.9609
F-statistic: 385.2 on 8 and 117 DF,  p-value: < 2.2e-16
```

If you run the LM of the same GLM final model you will see that the outputs were
exactly the same, it also gives the additional information that 96.1% of the variation

was explained by the model, even though we have removed interactions and a main effect.

```
# Plot full data and final model data
ggplot(data20, aes(x = Concentration, y = Time.Taken)) +
        theme_bw() + facet_wrap(~ Operator + Machine) +
        stat_boxplot(geom = "errorbar") + geom_boxplot()
```

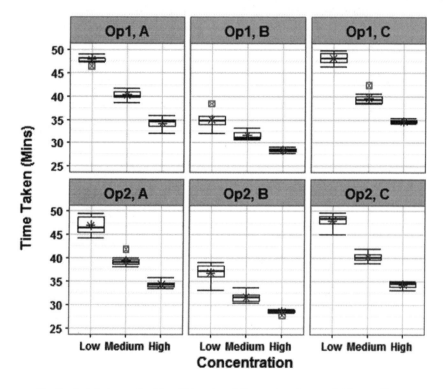

The final plot shows the Time.Taken by all the explanatory variables, as there was no difference between the Operators there is no need to include that in the plot. However it's still a good idea for visualization of the data plus we have already plotted the equivalent box plot for our final GLM model earlier.

Once the data has been input, you need to make sure the categorical explanatory variables are in the order you want. By redefining them as factors and using "*levels =*," you can order the levels as you wish.

You then assign a model to the data and name it, in this case "mod." The **glm( )** stands for the generalized linear model, and the fact that there is no distribution defined means it will treat the distribution as Gaussian or normal. The response variable is specified followed by the explanatory variables and interactions, then the data set to be used is included.

The next few commands are carried out on this "mod" object or "mod5" object, which was our simplified model, so the **summary( )** gives a summary of the model output. The summary can show us that the explanatory variables had an effect on "Time.Taken," but not where the differences were, so **lsmeans( )** does that by specifying the model along with which explanatory variable is of interest: "Machine", "Concentration", and then the interaction "Machine*Concentration." It also was used to calculate the confidence intervals for each factor level. A box plot was drawn to aid in interpreting the **lsmeans( )** output.

Confidence intervals were calculated on the estimates using the **confint( )** command. An ANOVA was run to test between all the models, and then the diagnostic plots were drawn.

The summary of the LM was run to highlight the similarities in the output.

Finally using **ggplot( )**, I plotted a box plot with the mean and statistical outliers clearly highlighted. As "Operator" was not significant I didn't need to include it on the plot, but did just to visualize the data fully.

The example also highlights that fitting a Gaussian GLM and a LM will produce the same summary results. It also shows the importance of interpreting the interaction term correctly, as although across all "Machines" the "High Concentration" produced the quickest "Time.Taken," then "Medium Concentration," then "Low Concentration," the gradient of that drop in "Time.Taken" depends on which "Machine" is being used.

Poisson GLM

A Poisson GLM concerns itself with count data, so the response variable will be counts and the explanatory variables can be either continuous or discrete. It can handle a few zero values, but if there are too many the model won't be a good fit and a zero-inflated Poisson model should be considered, an example is given later. In addition the mean and the variance should be close, if the variance is larger than the mean then there is over dispersion and a Poisson GLM will not be the best model to use, – see Negative Binomial GLM section.

The key thing to note when fitting a Poisson GLM is that the values given are on the log scale and will need to be transformed using the exponential to be given on the original response scale. Due to this fact sometimes this model is referred to as a log-linear model.

Example 7.4 investigates whether the income salary band people are in "Income" has any relationship with the number of children they have "Children."

EXAMPLE 7.4

We have a response variable of Children and an explanatory variable of Income. The response variable is discrete, counts, and the explanatory variable is also discrete. We are looking to see if Income has a relationship with the number of children the family has, Children.

```
# Input the data
Children = c(1,1,1,1,2,1,0,1,0,0,2,0,0,2,0,0,0,1,0,0,0,2,1,1,3,1,0,2,
             1,0,0,0,1,0,1,2,0,3,0,1,0,2,1,2,1,2,1,0,1,1,3,2,1,3,3,0,3,
             3,2,3,4,0,2,3,2,2,4,2,2,4,4,2,4,3,1,2,2,2,4,4,0,1,1,2,3,1,
             1,2,1,2,1,3,1,1,2,2,4,1,2,2,4,3,4,1,3,4,3,3,5,2,1,1,2,2,3,
             2,3,2,4,2,4,4,3,3,2,2,4,3,4,3,3,2,3,2,3,3,4,2,1,2,3,1,3,2,
             3,2,3,2,0,2)
Income = rep(c(">$50k","$25k-$50k","<$25k"), each = 50)
data21 = data.frame(Children, Income)

# Order the levels of the Income variable
data21$Income = factor(data21$Income, levels = c(">$50k",
    "$25k-$50k", "<$25k"))
```

R will order the levels of a variable "alphabetically"; this would mean Income would be $25k-$50k, <$25k, then >$50k which clearly is not the order we require. By using the command above we have reordered the Income explanatory variable appropriately.

```
# Look at the mean and variance of the Income groups
tapply(data21$Children, data21$Income, mean)
```

```
>$50k   $25k-$50k   <$25k
 0.86       2.18      2.64
```

```
tapply(data21$Children, data21$Income, var)
```

```
  >$50k     $25k-$50k      <$25k
0.735102    1.293469    1.051429
```

The variances are quite similar to the means with the exception of Income <$25k, differences of 0.12, 0.89, and 1.59, but more important the variances are not larger than the means, they are all smaller.

```
# Fit the full GLM and print the output
mod = glm(Children ~ Income, data = data21, family = poisson)
summary(mod)
```

```
Call:
glm(formula = Children ~ Income, family = poisson, data = data21)
```

```
Deviance Residuals:
    Min       1Q   Median       3Q      Max
-2.2978  -0.8952   0.1471   0.5251   1.7935
```

Coefficients:

	Estimate	Std. Error	z value	Pr(>|z|)	
(Intercept)	-0.1508	0.1525	-0.989	0.323	
Income$25k-$50k	0.9301	0.1801	5.165	2.40e-07	***
Income<$25k	1.1216	0.1756	6.388	1.68e-10	***

```
---
Signif. codes:  0 '***' 0.001 '**' 0.01 '*' 0.05 '.' 0.1 ' ' 1
```

(Dispersion parameter for poisson family taken to be 1)

```
    Null deviance: 159.03  on 149   degrees of freedom
Residual deviance: 108.39  on 147   degrees of freedom
AIC: 445.54
```

Number of Fisher Scoring iterations: 5

The output of the Poisson GLM showed that Income was significant to the model. The residual deviance (108.39) was also a lot lower than the null deviance (159.03), which suggests this model is better than the null model.

By looking at the estimates of the GLM, and back transforming using the exponential, we can see that the average number of Children for an Income of >$50k is exp(-0.1508) = 0.86, the average number of Children for an Income of $25k–$50k is exp(-0.1508 + 0.9301) = 2.18, and the average number of Children for an Income of <$25k is exp(-0.1508 + 1.1216) = 2.64. When back transforming the values, all addition needs to be carried out before using the exponential calculation, rather than using the exponentially transformed values and adding those values. If we were rounding that would give us 1, 2, and 3 Children for an Income of >$50k, $25k-$50k, and <$25k, respectively.

You can see that the estimate values here for number of children per income band are the same as the summary statistics calculated—the means. This is due to the fact that this is only a simple model, however the estimates should always be the values used as in more complex models they may differ from the summary statistics.

```
# Calculate confidence intervals for the estimates
confint(mod)
```

	2.5 %	97.5 %
(Intercept)	-0.4653798	0.1338905
Income$25k-$50k	0.5852679	1.2931358
Income<$25k	0.7866078	1.4767534

If we wanted confidence intervals on the estimates we can use the earlier values.

```
# Check for differences between all Incomes
lsmeans(mod, pairwise ~ Income)
```

```
$lsmeans
 Income        lsmean          SE  df   asymp.LCL  asymp.UCL
  >$50k    -0.1508229  0.15249819  NA  -0.4497138  0.1480681
$25k-$50k   0.7793249  0.09578263  NA   0.5915944  0.9670554
  <$25k     0.9707789  0.08703883  NA   0.8001859  1.1413719
```

Results are given on the log (not the response) scale.
Confidence level used: 0.95

```
$contrasts
           contrast    estimate         SE  df  z.ratio  p.value
>$50k - $25k-$50k     -0.9301478  0.1800833  NA   -5.165   <.0001
      >$50k - <$25k   -1.1216018  0.1755889  NA   -6.388   <.0001
  $25k-$50k - <$25k   -0.1914540  0.1294221  NA   -1.479   0.3009
```

Results are given on the log (not the response) scale.
P-value adjustment: tukey method for comparing a family of 3
estimates
Tests are performed on the log scale

The least-squares means results are also on the log scale, so these will need to be back transformed to be on the response scale. Those with an Income of <$50k had significantly fewer Children than both those with an Income of $25k–$50k and an Income of <$25k at the 99% confidence level. There was no significant difference in the number of Children between those with an Income of $25k-$50k and those with an Income of <$25k, p-value of 0.30.

If we wanted confidence intervals on each group level, we could use the top section of the output. So the confidence interval values, once exponentially transformed, we would get are 0.64 to 1.16, 1.81 to 2.63, and 2.23 to 3.13, respectively or rounding 1, 2, to 3, and also 2 to 3 for an Income of >$50k, $25k-$50k, and <$25k.

```
# Check the goodness of fit
1 - pchisq(summary(mod)$deviance, summary(mod)$df.residual)
```

[1] 0.9927758

As the distribution is not Gaussian the diagnostic plots do not need to be run, to test the goodness of fit for Poisson GLMs, the earlier command is used. The model fit was tested and the result showed that the model wasn't a bad fit, as $0.993 > 0.05$; we can never directly say it is a good fit from this test.

```
# Plot data
ggplot(data21, aes(x = Children)) + theme_bw() +
       facet_wrap(~Income) + geom_bar()
```

Number of Children by Income Band

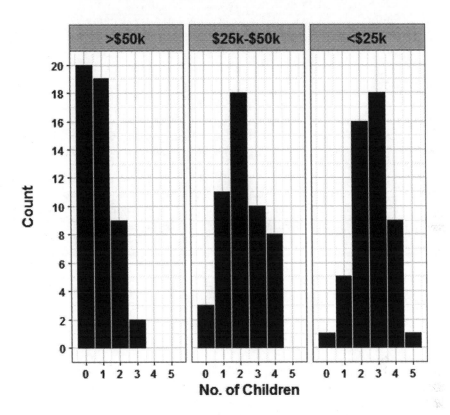

A bar chart is drawn to show the number of Children by the different Income salary bands, it does help to highlight the different shape of the >$50k Income compared to the other two.

The ***tapply( )*** command can be used to calculate summary statistics, such as the mean and variance, for different groups without too much hassle.

In this example the ***glm( )*** still stands for the generalized linear model, but this time the distribution is defined in the "***family = poisson***" section.

The rest follows the same processes as before with the only differences being that as the results are on the log scale they need to be back transformed using the exponential to give the response scale, and that the goodness of fit tests, or model fit, are slightly different

As the response variable is discrete, count data, a bar chart is drawn to show the data visually.

Make sure you check the model fit as if the data is over dispersed and a Poisson GLM is used instead of a negative binomial GLM, you increase the chance of detecting a significant difference between treatments when there shouldn't be one, which increases the likelihood of a Type I Error.

Negative Binomial GLM

A Poisson GLM may not always be appropriate with count data, if there aren't too many zeros and the data fails the Poisson GLM goodness of fit check, then you can try a negative binomial GLM (NBGLM). In general, NBGLMs are used when over dispersion in count data occurs, that is, when the variance is much larger than the mean.

Example 7.5 looks at the total fish caught in a 3-hour period "Fish.Caught" and whether the fact that tourists or locals "Group" caught them had an effect on the total count. To run a NBGLM, the R package **MASS** is required.

EXAMPLE 7.5

We have a response variable of Fish.Caught and an explanatory variable of Group. The response variable is discrete, counts, and the explanatory is also discrete. We are looking to see if Group has an effect on Fish.Caught.

```
# Input the data
Fish.Caught = c(0,3,3,0,4,8,6,1,2,1,0,1,1,2,4,1,3,3,4,3,1,3,1,2,2,3,8,5,
                2,2,4,2,2,5,3,2,0,4,3,1,5,0,1,4,1,2,2,2,0,3,2,9,9,1,5,7,2,
                4,6,8,1,4,2,16,10,11,3,5,12,11,1,0,5,2,3,8,1,7,5,10,13,4,
                10,1,0,2,7,7,3,1,9,4,2,2,2,1,10,2,9,2)
Group = rep(c("Tourists","Locals"), each = 50)
data22 = data.frame(Fish.Caught, Group)

# Look at the mean and variance of the Groups
tapply(data22$Fish.Caught, data22$Group, mean)

Locals  Tourists
 5.22     2.50

tapply(data22$Fish.Caught, data22$Group, var)

   Locals   Tourists
15.726122  3.479592
```

The variances are not similar to the means, especially for Locals, with differences of 10.51 and 0.98. More important the variances are larger than the means that suggests over dispersion; we will fit a Poisson GLM to verify.

```
# Fit a Poisson GLM and check the goodness of fit
mod = glm(Fish.Caught ~ Group, data = data22, family = poisson)
1 - pchisq(summary(mod)$deviance, summary(mod)$df.residual)
```

[1] 3.455791e-12

The goodness of fit test for the Poisson GLM showed that the model was not a good fit, p-value < 0.05, which suggests over dispersion in correlation with the above information.

```
# Fit a Negative Binomial GLM and print the output, and check the
goodness of fit
library(MASS)
mod2 = glm.nb(Fish.Caught ~ Group, data = data22)
summary(mod2)
```

```
Call:
glm.nb(formula = Fish.Caught ~ Group, data = data22,
        init.theta = 3.008476736, link = log)

Deviance Residuals:
    Min       1Q    Median       3Q      Max
-2.4605   -0.9010   -0.2462   0.6199   2.0030

Coefficients:
               Estimate  Std. Error  z value  Pr(>|z|)
(Intercept)      1.6525      0.1024   16.143   < 2e-16 ***
GroupTourists   -0.7362      0.1585   -4.644  3.41e-06 ***
---
Signif. codes:  0 '***' 0.001 '**' 0.01 '*' 0.05 '.' 0.1 ' ' 1

(Dispersion parameter for Negative Binomial (3.0085) family taken to
be 1)

Null deviance: 128.35  on 99  degrees of freedom
Residual deviance: 106.50  on 98  degrees of freedom
AIC: 467.68

Number of Fisher Scoring iterations: 1

              Theta:     3.008
          Std. Err.:     0.791

2 x log-likelihood:   -461.683
```

The output of the NBGLM showed that Group was significant to the model at the 99% confidence level. The residual deviance (106.50) was also slightly lower than the null deviance (128.35) that suggests this model is better than the null model. Theta represents the shape parameter of the distribution, so the dispersion is 3.01. The closer theta is to 0 the more dispersed the data and the larger theta is the closer it is to a Poisson model.

By looking at the estimates and back transforming using the exponential, we can see that the number of Fish.Caught for Locals is exp(1.6525) = 5.22 and the number of Fish.Caught for Tourists is exp(1.6525 − 0.7362) = 2.50. If we were rounding that would give us 5 and 3 Fish.Caught for Locals and Tourists, respectively.

```
# Calculate estimate confidence intervals
confint(mod2)
```

```
                2.5 %      97.5 %
(Intercept)     1.453493   1.8551163
GroupTourists  -1.048322  -0.4266079
```

If we wanted confidence intervals on the estimates we could use the earlier values.

```
# Calculate confidence intervals - only first section of output shown
lsmeans(mod2, pairwise ~ Group)
```

```
$lsmeans
  Group    lsmean         SE  df  asymp.LCL  asymp.UCL
 Locals  1.6524974  0.1023685  NA  1.4518589   1.853136
Tourists  0.9162907  0.1210284  NA  0.6790794   1.153502
```

```
Results are given on the log (not the response) scale.
Confidence level used: 0.95
```

If we wanted confidence intervals on each group level, we could use the output from a least-squares means that includes the interaction term. So the confidence interval values, once exponentially transformed, we would get are 4.27 to 6.38, and 1.97 to 3.17, respectively or rounding 4 to 6, and 2 to 3 for Locals and Tourists.

```
# Check the goodness of fit
1 - pchisq(summary(mod2)$deviance, summary(mod2)$df.residual)
```

```
[1] 0.2618407
```

The model fit was tested and the result showed that the model wasn't a bad fit, as 0.262 > 0.05, we can never directly say it is a good fit from this test.

```
# Plot data
ggplot(data22, aes(x = Fish.Caught)) + theme_bw() +
       facet_wrap(~Group) + geom_bar()
```

Number of Fish Caught within 3 Hours by Group

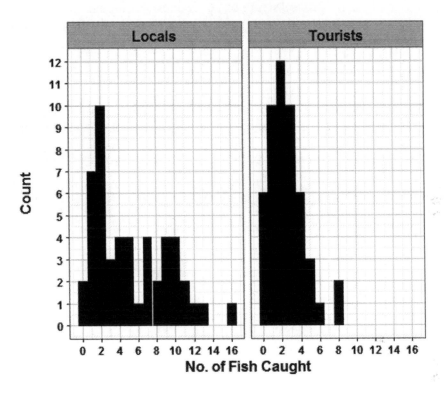

A bar chart is drawn to show the number of Fish.Caight by the different Groups, Locals and Tourists, it does help to highlight the different shapes of the two distributions and highlights the over dispersion in the Locals Group.

In this example the **glm.nb()** stands for the NBGLM. The rest then follows the same processes as with the Poisson GLM example.

The main things to remember when fitting a NBGLM is to look at the mean and variance of the data, then fit a Poisson GLM before the NBGLM, and also that the values given are also on the log scale and will need to be transformed using the exponential for the response scale.

Binomial GLM

A binomial GLM is used when the response variable is binary data. The explanatory variables can be either discrete or continuous. The model doesn't cope very well if there is a clear separation of zeros and ones in different groups, in which case a bias-reduction binomial-response GLM should be used, there is an example given later.

The key thing to note when fitting a binomial GLM is that the values given are on the log odds scale and will need to be transformed using the exponential to be given on the odds ratio scale. Due to this fact sometimes this model is referred to as a logistic regression model or a logit model. When the odds ratio of A to B is less than 1 it is easier to interpret if it is swapped B to A by dividing 1 by the odds ratio that is less than 1, then the odds can also be stated.

Example 7.6 investigates whether the device used "Device" or the three types of terrain covered "Terrain" has an effect on whether a detection was indicated or not "Detection."

EXAMPLE 7.6

We have a response variable of Detection and two explanatory variables of Device and Terrain. The response variable is discrete, binary, and both explanatory variables are also discrete. We are looking to see if Device, Terrain, or the interaction between the two has an effect on Detection.

```
# Input the data
Detection = c(1,1,0,1,1,0,1,1,1,1,1,1,0,1,1,0,1,1,1,1,1,0,1,1,1,1,1,
              1,1,1,0,1,0,0,0,0,1,0,0,1,0,1,0,1,0,1,1,0,1,0,1,1,0,0,1,0,
              0,1,1,0,0,0,0,1,0,0,0,0,0,0,0,0,0,1,0,0,0,1,0,0,0,0,0,0,0,
              0,0,0,0)
Device = rep(c("A","B"), each = 15,3)
Terrain = rep(c("Flat","Bumpy","Marsh"), each = 30)
data23 = data.frame(Detection, Terrain, Device)

# Order the levels of the Terrain variable
data23$Terrain = factor(data23$Terrain,
        levels = c("Flat","Bumpy","Marsh"))

# Fit the full Binomial GLM model and print the output
mod = glm(Detection ~ Terrain*Device, data = data23,
        family = binomial)
summary(mod)

Call:
glm(formula = Detection ~ Terrain * Device, family = binomial,
        data = data23)

Deviance Residuals:
    Min      1Q   Median      3Q     Max
-2.0074  -0.5350  -0.3715  0.6681  2.3272
```

```
Coefficients:
                     Estimate  Std. Error  z value  Pr(>|z|)
(Intercept)           1.38629     0.64550    2.148   0.03174 *
TerrainBumpy         -1.79176     0.83333   -2.150   0.03155 *
TerrainMarsh         -3.25810     0.99679   -3.269   0.00108 **
DeviceB               0.48551     0.99679    0.487   0.62621
TerrainBumpy:DeviceB  0.05349     1.24065    0.043   0.96561
TerrainMarsh:DeviceB -1.25276     1.62540   -0.771   0.44086
---
Signif. codes:  0 '***' 0.001 '**' 0.01 '*' 0.05 '.' 0.1 ' ' 1

(Dispersion parameter for binomial family taken to be 1)

Null deviance: 124.366  on 89  degrees of freedom
Residual deviance:  86.838  on 84  degrees of freedom
AIC:  98.838

Number of Fisher Scoring iterations: 5
```

The output of the full binomial GLM that contained all the main effects and interaction showed that the first term to remove was the two-way interaction.

```
# Simplify the model by removing one interaction or explanatory variable
# at a time
# Output omitted for space saving - except for the final model
mod2 = glm(Detection ~ Terrain + Device, family = binomial,
       data = data23)
summary(mod2)
```

When the model was refitted it then showed that the Device variable could be removed, p-value of 0.598.

```
mod3 = glm(Detection ~ Terrain, family = binomial, data = data23)
summary(mod3)

Call:
glm(formula = Detection ~ Terrain, family = binomial, data = data23)

Deviance Residuals:
    Min       1Q    Median       3Q      Max
-1.8930  -0.4590   -0.4590   0.6039   2.1460

Coefficients:
              Estimate  Std. Error  z value  Pr(>|z|)
(Intercept)     1.6094      0.4899    3.285   0.00102 **
TerrainBumpy   -1.7430      0.6115   -2.850   0.00437 **
TerrainMarsh   -3.8067      0.7812   -4.873   1.1e-06 ***
---
Signif. codes:  0 '***' 0.001 '**' 0.01 '*' 0.05 '.' 0.1 ' ' 1

(Dispersion parameter for binomial family taken to be 1)

Null deviance: 124.366  on 89  degrees of freedom
Residual deviance:  87.994  on 87  degrees of freedom
AIC:  93.994

Number of Fisher Scoring iterations: 4
```

The final output of the binomial GLM showed that Terrain was significant to the model. The residual deviance (87.99) was lower than the null deviance (124.37) that suggests this model is better than the null model.

By looking at the estimates of the binomial GLM we can see that when there is Bumpy Terrain instead of Flat Terrain, the log odds of Detection decreases by 1.74, so therefore the odds ratio of Detection is exp(-1.74) = 0.175. In other words for Flat Terrain instead of Bumpy Terrain the odds of Detection is almost 6 times greater, (1/0.175) = 5.71.

For Marsh Terrain instead of Flat Terrain, the log odds of Detection decreases by 3.81, so therefore the odds ratio of Detection is exp(-3.81) = 0.022. In other words for Flat Terrain instead of Marsh Terrain the odds of Detection is 45 times greater, (1/0.022) = 45.00.

For Marsh Terrain instead of Bumpy Terrain, the log odds of Detection decreases by (-3.81 − (-1.74)) = -2.06, so therefore the odds ratio of Detection is exp(-2.06) = 0.127, in other words for Bumpy Terrain instead of Marsh Terrain the odds of Detection is almost 8 times greater, (1/0.127) = 7.88.

```
# Calculate confidence intervals
confint(mod3)
```

```
                    2.5 %       97.5 %
(Intercept)      0.7327236    2.6955982
TerrainBumpy    -3.0280653   -0.5979537
TerrainMarsh    -5.5264219   -2.4067335
```

The confidence intervals for the differences are also calculated on the log-likelihood scale, so if we wanted confidence intervals on the earlier values, we would have to transform them using the exponential. The confidence interval values we would get are 0.048 to 0.550 and 0.004 to 0.090, which is the equivalent of the odds of Detections on Flat Terrain being 2 times to 21 times greater than on Bumpy Terrain, and the odds of Detections on Flat Terrain being 11 times to 251 times greater than on Marsh Terrain.

To calculate the difference between Bumpy Terrain and Marsh Terrain, you need to subtract the values again, so the log odds confidence interval would be (-5.53 − (-3.03)) = -2.50 to (-2.41 − (-0.60)) = -1.81. This would make the interval for the odds ratio 0.082 to 0.164, which is the equivalent of the odds of Detections on Bumpy Terrain being 6 to 12 times greater than on Marsh Terrain.

```
# Check for differences between all Terrains
lsmeans(mod3, pairwise ~ Terrain)
```

```
$lsmeans
Terrain       lsmean         SE  df    asymp.LCL   asymp.UCL
   Flat    1.6094379  0.4898979  NA    0.6492556    2.569620
  Bumpy   -0.1335314  0.3659625  NA   -0.8508048    0.583742
  Marsh   -2.1972246  0.6085305  NA   -3.3899224   -1.004527
```

Results are given on the logit (not the response) scale.
Confidence level used: 0.95

```
$contrasts
     contrast  estimate        SE  df  z.ratio  p.value
  Flat - Bumpy  1.742969  0.6114970  NA    2.850   0.0122
  Flat - Marsh  3.806662  0.7812230  NA    4.873   <.0001
 Bumpy - Marsh  2.063693  0.7100971  NA    2.906   0.0102
```

Results are given on the log (not the response) scale.
P-value adjustment: tukey method for comparing a family of 3
estimates
Tests are performed on the log scale

The lsmeans results are also on the log-likelihood scale, so these will need to be back transformed to be on the response scale. The log odds of getting a Detection was significantly greater on the Flat Terrain compared to both the Bumpy and Marsh Terrain at the 95% and 99% confidence level, respectively. Also the log odds of getting a Detection was significantly greater on the Bumpy Terrain compared to the Marsh Terrain at the 99% confidence level.

```
# Check the goodness of fit
1 - pchisq(summary(mod3)$deviance, summary(mod3)$df.residual)
```

[1] 0.4500269

The model fit was tested and the result showed that the model wasn't a bad fit, as $0.450 > 0.05$. However, we can never directly say it is a good fit from this test.

```
# Compare the models
anova(mod, mod2, mod3, test = "Chisq")
```

Analysis of Deviance Table

```
Model 1: Detection ~ Terrain * Device
Model 2: Detection ~ Terrain + Device
Model 3: Detection ~ Terrain
  Resid. Df  Resid. Dev  Df  Deviance  Pr(>Chi)
1        84      86.838
2        86      87.715  -2  -0.87609    0.6453
3        87      87.994  -1  -0.27947    0.5970
```

To check the model could be simplified an ANOVA comparing all the models was run using the chi-square test instead of the *F* test due to non-normality. All these p-values suggest no significant difference between the models; hence it is fine to use the simpler model.

```
# Create table of counts and plot data
Table3 = as.table(ftable(Detection ~ Terrain + Device, data = data23))
data24 = as.data.frame(Table3)
ggplot(data24, aes(x = Terrain, y = Freq, fill = Detection)) +
    theme_bw() + facet_wrap(~Device) +
    geom_bar(stat = "identity", position = "dodge")
```

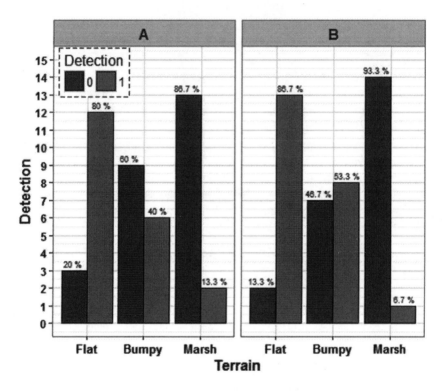

Frequency of Detections by Terrain and Device

Due to the setup of the data it is easier to create a table of results for plotting purposes. A bar chart is drawn to show the number of Detections by the different Terrains and the two Devices. As Device was not significant to the model it doesn't need to be included in the plot, however for visualization purposes I have included it. The graph does help to highlight the differences in Detections between the three Terrains as well as show the similarities across Devices.

In this example the **glm()** still stands for the generalized linear model, but this time the distribution is defined in the "**family = binomial**" section.

The rest follows the same processes as previous GLM examples with the only differences being that as the results are on the log odds scale they need to be back transformed using the exponential to give the odds ratio scale, and that the goodness of fit tests, or model fit, are the same as those for the Poisson GLM and NBGLM.

Bias-Reduction Binomial-Response RGLM

A bias-reduction binomial-response (BRGLM) should be used when the response is binary data but there is either a clear separation of zeros and ones in different groups; that is, Group A has all zeros and Group B has all ones, or mostly so. The same may be true when the response is continuous, but isn't always necessarily so; however it is always worth comparing the Binomial GLM and BRGLM if you are unsure.

As with the binomial GLM, the results will be on the log odds scale and will therefore need to be transformed to the odds ratio scale using the exponential transformation.

Example 7.7 investigates whether the concentration "Concentration" has an effect on whether there was a detection indicated or not "Detection." To fit a BRGLM the R package required is called **brglm**; another package that is needed for parts of this example is **MASS**.

EXAMPLE 7.7

We have a response variable of Detection and an explanatory variable of Concentration. The response variable is discrete, binary, and although the explanatory was recorded as a discrete variable, it is classed as continuous for the analysis. It makes sense to be able to have Concentrations between those recorded and assume it follows the same pattern. We are looking to see if Concentration has an effect on Detection.

```
# Input the data
Detection = c(0,0,0,0,0,0,0,0,0,0,0,0,0,0,0,0,0,0,0,0,0,0,0,0,0,0,0,0,0,0,
              0,0,1,1,0,0,0,0,1,1,0,0,0,0,1,1,1,0,0,0,1,1,1,0,0,0,1,1,1,
              1,0,0,1,1,1,0,0,0,1,1,1,1,0,0,1,1,1,1,1,1,1,1,1,1,1,1,1,1,
              1,1,1,1,1,1,1,1,1,1)
Concentration = rep(c(100,120,140,160,180,200,220,240,260,280,300,
         320,340,360,380,400), each = 6)
data25 = data.frame(Detection, Concentration)

# Fit a Binomial GLM and print the output
library(brglm)
mod = brglm(Detection ~ Concentration, family = binomial,
        data = data25)
summary(mod)

Call:
brglm(formula = Detection ~ Concentration, family = binomial,
        data = data25)

Coefficients:
                Estimate  Std. Error  z value  Pr(>|z|)
(Intercept)     -7.106552    1.375278   -5.167  2.37e-07 ***
Concentration    0.027312    0.005176    5.277  1.31e-07 ***
---
```

```
Signif. codes:  0 '***' 0.001 '**' 0.01 '*' 0.05 '.' 0.1 ' ' 1
```

(Dispersion parameter for binomial family taken to be 1)

```
Null deviance: 124.397  on 95  degrees of freedom
Residual deviance:  65.611  on 94  degrees of freedom
Penalized deviance: 52.71343
AIC:  69.611
```

The output of the BRGLM showed that Concentration was significant to the model at the 99% confidence level. The residual deviance (65.61) and the penalized deviance (52.71) also were around half the null deviance (124.40) that suggests this model is better than the null model.

By looking at the estimates we can see that for a one unit increase in Concentration, the log odds of Detection increases by 0.027, so therefore the odds ratio of detection is 1.0277.

```
# Calculate confidence intervals for the estimates
confint(mod)
```

```
                    2.5 %        97.5 %
(Intercept)    -10.84743941  -4.72744612
Concentration    0.01835369   0.04144744
```

The confidence intervals also are calculated on the log-likelihood scale, so if we wanted confidence intervals on the Concentration estimate we would have to transform it using the exponential. The confidence interval values we would get are 1.019 to 1.042.

```
# Check the goodness of fit
1 - pchisq(summary(mod)$deviance, summary(mod)$df.residual)
```

```
[1] 0.9885665
```

The model fit was tested and the result showed that the model wasn't a bad fit, as $0.989 > 0.05$; we can never directly say it is a good fit from this test.

```
# Calculate concentrations for given probabilities of detection
library(MASS)
dose.p(mod, p = c(0.5, 0.75, 0.95, 0.99))
```

```
              Dose        SE
p = 0.50:  260.1943  11.19369
p = 0.75:  300.4182  13.66244
p = 0.95:  368.0000  23.48199
p = 0.99:  428.4368  33.99168
```

We can calculate the required Concentration, termed "Dose" in the R code, for given probabilities of detection (PDet) using the fitted model. We can see that if we want a PDet of 50% the Concentration required is 260.19 whereas if we need a PDet of 99% the Concentration required is 428.44.

```
# Predict values using the model and create 95% confidence intervals
# around the Concentration values
xcis = dose.p(mod, p = seq(0.01, 0.99, 0.01))
tempv50 = data.frame(prop = seq(0.01,0.99,0.01),
        v50 = as.numeric(xcis), v50.se = as.numeric(attr(xcis,"SE")))
tempv50$lowerv50 = tempv50$v50 - (1.96*(tempv50$v50.se))
tempv50$upperv50 = tempv50$v50 + (1.96*(tempv50$v50.se))
tempConc = seq(80, 420, 1)
tmpModFit = data.frame(Concentration = tempConc,
Proportion = predict(mod, newdata = data.frame(
Concentration = tempConc), type = "response"))
```

This section of the code sets up a new data set then predicts new Concentration values from the probability of detections, from 0.01 to 0.99 by steps of 0.01. The predictions are calculated given the fit of the BRGLM, so we can plot a "line of best fit" on the data along with 95% confidence intervals around those new Concentration values.

```
# Plot data with line of best fit and confidence intervals
ggplot(data25, aes(Concentration, Detection)) +
        geom_point() + theme_bw() +
        geom_line(aes(Concentration, Proportion), tmpModFit) +
        geom_line(aes(lowerv50, prop), tempv50) +
        geom_line(aes(upperv50, prop), tempv50)
```

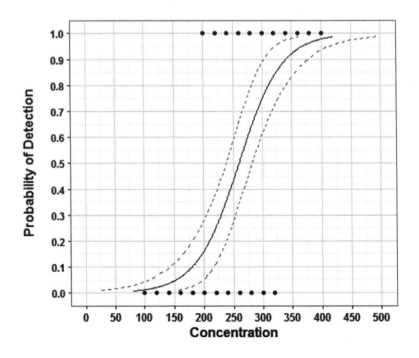

A bar chart of these results or a plain scatterplot of PDet vs Concentration would not be very useful, so we create a scatterplot that includes the BRGLM fitted model along with 95% confidence intervals around the Concentration values. This highlights the curve of the model and we can see roughly what Concentration is required for given PDets or we can use the dose.p() command.

In this example the ***brglm( )*** stands for the BRGLM.

The rest follows the same processes as the binomial GLM example and as before the model output results are on the log odds scale, so they need to be back transformed using the exponential to give the odds ratio scale. The only difference between the examples, in terms of output, is that as the explanatory variable is continuous in the BRGLM we can overlay a fitted curve to the plot and predict which "Concentrations" are required for given PDets.

The key thing to remember when deciding between using a binomial GLM and a BRGLM is the separation of zeros and ones. In this example we had a few lower "Concentrations" that were all zeros, and a few higher "Concentrations" that were all ones, hence fitting a BRGLM instead of a binomial GLM.

Zero-Inflated Models

In some cases you will have count data that may include a lot of zeros, and fitting a Poisson GLM or a NBGLM to this data may give incorrect results due to this. It's used when there are two processes occurring, for example the chance of seeing a shooting star is one process, and if they can be seen, how many can be seen is the second process.

You wouldn't want to exclude the zeros from the data so you can fit one of the previous count models, and likewise you wouldn't want to just convert the count data into binary data to run a binomial GLM or a BRGLM. This is where the zero-inflated models come into play as they will fit both types of models, binary and count, to the data to account for the large number of zeros.

When thinking about fitting zero-inflated models always try the Poisson GLM and NBGLM first to assess their goodness of fit. If they don't fit the data, then try a zero-inflated Poisson (ZIP) model, double checking there are more zeros in the data than would be expected from a regular Poisson GLM. Consider the zero-inflated negative binomial (ZINB) model as well; if the dispersion parameter, theta, is significant then choose the ZINB model over the ZIP model, but if it's not significant then stick with the ZIP model.

Note that as with the Poisson GLMs and NBGLMs, the count results from the ZIP and ZINB models will be given on the log scale and will need to be transformed using the exponential to be given on the original response scale. In addition, the zero-inflation section replicates a binomial GLM, so the results will be given on the log odds scale and will need to be transformed using the exponential to be given on the odds ratio scale.

To compare zero-inflated models to their count data equivalents we cannot use the ANOVA test as we have done before as this test only works when models are nested (i.e., a model with two main effects is nested within a model with the two main effects and the interaction between them). To compare a zero-inflated model to its counterpart we use Vuong's non-nested hypothesis test. However you can only compare a Poisson GLM to a ZIP and a NBGLM to a ZINB, you cannot use it to compare a NBGLM to a ZIP or a Poisson GLM to a ZINB.

Example 7.8 investigates whether the location, City or Countryside, people live in "Location" has any relationship with the number of shooting stars they saw in a period of an hour "Shooting.Stars." To run a zero-inflated model in R the package *pscl* is required.

EXAMPLE 7.8

We have a response variable of Shooting.Stars and an explanatory variable of Location. The response variable is discrete, counts, and the explanatory variable is also discrete. We are looking to see if Location has a relationship with the number of shooting stars seen in an hour—Shooting.Stars.

```
# Input the data
Shooting.Stars = c(3,0,4,3,0,1,4,2,0,2,0,1,1,1,0,3,0,0,0,0,3,0,3,4,0,0,4,
                   2,3,0,2,1,3,2,1,2,2,2,0,0,0,0,0,3,1,1,4,5,0,2,2,5,4,6,4,
                   5,5,7,6,0,4,6,6,0,5,4,4,3,0,6,4,2,5,1,4,5,6,7,0,7,4,4,
                   6,6,2,0,5,7,1,4,3,7,7,3,7,7,7,7,4,5,4)
Location = rep(c("City", "Countryside"), each = 50)
data26 = data.frame(Shooting.Stars, Location)

# Fit a Poisson GLM and check the goodness of fit
mod = glm(Shooting.Stars ~ Location, data = data26,
          family = poisson)
1 - pchisq(summary(mod)$deviance, summary(mod)$df.residual)
```

[1] 7.415786e-05

```
# Fit a Negative-Binomial GLM and check the goodness of fit
library(MASS)
mod2 = glm.nb(Shooting.Stars ~ Location, data = data26)
1 - pchisq(summary(mod2)$deviance, summary(mod2)$df.residual)
```

[1] 0.002797011

Both the Poisson GLM and the NBGLM were not good fits for the data, both p-values < 0.05.

```
# Load library, fit a zero-inflated Poisson model and print the
output
library(pscl)
mod3 = zeroinfl(Shooting.Stars ~ Location, data = data26)
summary(mod3)

Call:
zeroinfl(formula = Shooting.Stars ~ Location, data = data26)

Pearson residuals:
     Min        1Q    Median        3Q       Max
-1.74379  -0.96316   0.03156   0.62335   2.22068

Count model coefficients (poisson with log link):
                      Estimate  Std. Error  z value  Pr(>|z|)
(Intercept)             0.7497      0.1352    5.544  2.96e-08 ***
LocationCountryside     0.8341      0.1516    5.501  3.77e-08 ***

Zero-inflation model coefficients (binomial with logit link):
                      Estimate  Std. Error  z value  Pr(>|z|)
(Intercept)            -0.9828      0.4152   -2.367    0.0179 *
LocationCountryside    -1.2940      0.6555   -1.974    0.0484 *
---
Signif. codes:  0 '***' 0.001 '**' 0.01 '*' 0.05 '.' 0.1 ' ' 1

Number of iterations in BFGS optimization: 9
Log-likelihood: -186.8 on 4 Df
```

The zero-inflated section of the ZIP model output showed that there were significantly more zeros than would be expected in a Poisson model at the 95% confidence level, p-value of 0.0179. It also showed that Location was significant to the number of zeros at the 95% confidence level, p-value of 0.048. The count section of the ZIP model showed that Location was significant to the number of Shooting.Stars at the 99% confidence level, p-value of 3.77e-08.

The results are given in logs for the count section and log odds for the zero-inflation section, so the average number of Shooting.Stars seen in the City when they can be seen is exp(0.7497) = 2.12, and the average number of Shooting.Stars seen in the Countryside is exp(0.7497 + 0.8341) = 4.87, or rounded is 2 and 5 for City and Countryside, respectively.

In addition, the log odds of not seeing any Shooting.Stars (seeing zero Shooting.Stars) in the Countryside compared to the City is -1.294, so therefore the odds ratio of not seeing any Shooting.Stars is exp(-1.2940) = 0.274. In other words for the City instead of the Countryside the odds of not seeing any Shooting.Stars is almost 4 times greater: (1/0.274) = 3.65.

```
# Calculate confidence intervals for the ZIP model estimates
confint(mod3)
```

	2.5 %	97.5 %
count_(Intercept)	0.4846500	1.014759264
count_LocationCountryside	0.5369426	1.131297025
zero_(Intercept)	-1.7965771	-0.168984653
zero_LocationCountryside	-2.5788261	-0.009143994

The confidence intervals for the estimates are also calculated on the log and the log-likelihood scale. We can use this output to calculate the confidence intervals around the difference between Countryside and City in terms of the odds of seeing a Shooting.Star, and we can use it for the confidence interval around the average number of Shooting.Stars seen in the City.

So the log odds confidence interval of the difference between Countryside and City, in terms of seeing a Shooting.Star is -2.579 to -0.009, so the odds ratio is 0.076 to 0.991. In other words, the odds of not seeing any Shooting.Stars in the City compared to the Countryside is 1 to 13 times greater. The confidence interval around the City average number of Shooting.Stars is between 2 and 3.

```
# Predict expected number of shooting stars per location from the
model
data27 = expand.grid(levels(data26$Location))
colnames(data27) = c("Location")
data27$Est.ShStars = predict(mod3, data27); data27
```

```
    Location  Est.ShStars
1       City     1.540001
2 Countryside    4.419998
```

So taking into account the zeros and the counts of Shooting.Stars, the predicted average number of Shooting.Stars seen in the City will be 1.54, or 2, and the predicted average number of Shooting.Stars seen in the Countryside will be 4.42, or 4.

```
# Compare Poisson model to zero-inflated Poisson model
vuong(mod, mod3)
```

```
Vuong Non-Nested Hypothesis Test-Statistic:
(test-statistic is asymptotically distributed N(0,1) under the
 null that the models are indistinguishible)
---------------------------------------------------------------
```

	Vuong z-statistic	H_A	p-value
Raw	-2.222958	model2 > model1	0.013109
AIC-corrected	-1.869046	model2 > model1	0.030808
BIC-corrected	-1.408046	model2 > model1	0.079559

Comparing the Poisson GLM against the ZIP model using the Vuong test showed that the ZIP model was better than the Poisson GLM as the p-value < 0.05. The null hypothesis here is that the models are the same or model 1 (Poisson GLM) is better, and the alternative hypothesis is that model 2 (ZIP) is better. However, we should still fit a ZINB to check whether that would be a better fit for our data.

```
# Fit a zero-inflated Negative Binomial model and print the results
mod4 = zeroinfl(Shooting.Stars ~ Location, dist = "negbin",
        data = data26)
summary(mod4)
```

```
Call:
zeroinfl(formula = Shooting.Stars ~ Location, dist = "negbin",
        data = data26)

Pearson residuals:
    Min       1Q    Median      3Q      Max
-1.74379  -0.96315  0.03156  0.62334  2.22068

Count model coefficients (negbin with log link):
                    Estimate  Std. Error  z value  Pr(>|z|)
(Intercept)           0.7497     0.1352     5.544  2.96e-08 ***
LocationCountryside   0.8341     0.1516     5.501  3.77e-08 ***
Log(theta)           12.9491    88.2420     0.147     0.883

Zero-inflation model coefficients (binomial with logit link):
                    Estimate  Std. Error  z value  Pr(>|z|)
(Intercept)          -0.9828     0.4152    -2.367    0.0179 *
LocationCountryside  -1.2940     0.6555    -1.974    0.0484 *
---
Signif. codes:  0 '***' 0.001 '**' 0.01 '*' 0.05 '.' 0.1 ' ' 1

Theta = 420450.1598
Number of iterations in BFGS optimization: 40
Log-likelihood: -186.8 on 5 Df
```

Checking the ZINB model shows that theta, the dispersion parameter, was not significant to the model, p-value of 0.883. This means that the ZIP model is better suited to the data than the ZINB model.

```
# Plot data
ggplot(data26, aes(x = Shooting.Stars)) + theme_bw() +
        facet_wrap(~Location) + geom_bar()
```

Number of Shooting Stars Seen in an Hour by Location

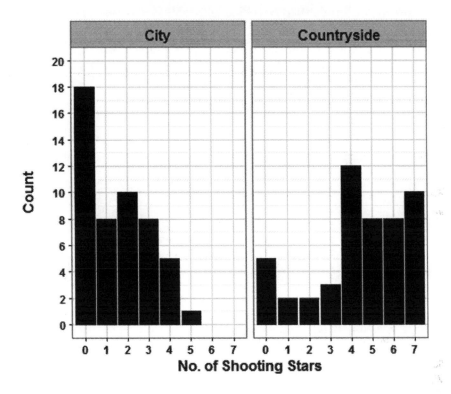

A bar chart is drawn to show the number of Shooting.Stars seen in an hour by the different Locations, City and Countryside; it does help to highlight the different shapes of the two distributions.

In this example the **zeroinfl()** stands for the zero-inflated model. If there is no distribution specified it assumes the Poisson, ZIP, but if the "**dist = negbin**", then it uses the negative binomial, ZINB.

The rest follows the same processes as previous Poisson GLM and NBGLM examples with the only differences being that the results are given in both the log scale and the log odds scale dependent on which section of the model output you are looking at. The goodness of fit test requires the zero-inflated model to be compared to the count equivalent using Vuong's hypothesis test.

The key thing to remember when fitting zero-inflated models is to verify that a Poisson GLM or NBGLM doesn't fit the data first, then to fit the ZIP, then to fit the ZINB and compare them.

Ordinal Logistic Regression

An ordinal logistic regression (OLR) model is used when the response variable is ordinal; the explanatory variables can be either discrete or continuous.

The key thing to note when fitting an OLR model is that, as with the binomial GLM, the values given are on the log odds scale and will need to be transformed using the exponential to be given on the odds ratio scale. Due to this fact sometimes this model is referred to as a proportional odds logistic regression model.

The OLR output will produce t-values instead of p-values. A general rule of thumb is that a t-value < -2 or a t-value > 2 means that the explanatory variable is significant to the model. There is a lot of discussion surrounding whether it is appropriate to calculate p-values from these t-values. However it can be done and I have included the R code to do so if you prefer.

A key assumption with OLR is that the steps between the response variable, for example, *strongly agree* to *agree*, *agree* to *disagree*, *disagree* to *strongly disagree*, are all equal or in other words, they are all linear.

Example 7.9 investigates whether the any of the five systems "System" or either of the groups of participants *"Group"* gave different Likert responses "Likert.Responses" in relation to whether the system performed *poorly* or *well* with their current equipment. There are two packages required to run the OLR model in R, **MASS** and **AER**. There is an additional package required to create the Likert plot called **HH**.

EXAMPLE 7.9

We have a response variable of Likert.Response and two explanatory variables of System and Group. The response variable is discrete, ordinal, and both the explanatory variables are discrete. We are looking to see if the System, Group, or the interaction between the two has a relationship with the Likert.Response of participants.

```
# Input the data
System = rep(c("A","B","C","D","E"), each = 4, 2)
Likert.Response = rep(c("VP","P","W","VW"), 10)
Group = rep(c("Group A","Group B"), each = 20)
Response = c(4,6,1,0,6,5,0,0,0,1,4,6,1,4,5,1,0,3,7,1,5,5,2,0,5,6,
             1,0,0,0,7,5,2,2,6,2,2,3,6,1)
Subjects = rep(c(11,12), each = 20)
data28 = data.frame(System, Likert.Response, Group, Response)

# Make sure variable is ordered correctly
data28$Likert.Response = factor(data28$Likert.Response,
        levels = c("VP","P","W","VW"),
        labels = c("Very Poorly", "Poorly", "Well", "Very Well"))
```

It is crucial to verify that the ordinal response scale is in the correct order, otherwise the model output will be meaningless.

```
# Fit the full OLR model and print the output
library(MASS)
mod = polr(Likert.Response ~ System*Group, weights = Response,
        data = data28)
summary(mod)
```

Re-fitting to get Hessian

```
Call:
polr(formula = Likert.Response ~ System * Group, data = data28,
        weights = Response)

Coefficients:
                            Value   Std. Error    t value
SystemB                 -0.719159       0.7985  -0.900674
SystemC                  4.704055       0.9417   4.995497
SystemD                  2.001315       0.8282   2.416463
SystemE                  2.678277       0.8380   3.196078
GroupGroup B            -0.003012       0.7803  -0.003861
SystemB:GroupGroup B     0.551690       1.1076   0.498091
SystemC:GroupGroup B    -0.280103       1.1257  -0.248815
SystemD:GroupGroup B     0.483906       1.1458   0.422347
SystemE:GroupGroup B    -0.696591       1.1253  -0.619018

Intercepts:
                        Value  Std. Error  t value
Very Poorly|Poorly    -0.3962      0.5599  -0.7077
Poorly|Well            1.7775      0.6083   2.9220
Well|Very Well         4.5998      0.7449   6.1754

Residual Deviance: 233.0714
AIC: 257.0714
```

The full OLR model showed that the interaction was not significant to the model.

```
# If you would prefer p-values use the code below
library(AER)
coeftest(mod)
```

Re-fitting to get Hessian

```
z test of coefficients:
                         Estimate  Std. Error  z value   Pr(>|z|)
SystemB                -0.7191594   0.7984678  -0.9007   0.367762
SystemC                 4.7040550   0.9416590   4.9955  5.868e-07 ***
SystemD                 2.0013154   0.8282002   2.4165   0.015672 *
SystemE                 2.6782767   0.8379884   3.1961   0.001393 **
GroupGroup B           -0.0030124   0.7803013  -0.0039   0.996920
SystemB:GroupGroup B    0.5516897   1.1076083   0.4981   0.618420
SystemC:GroupGroup B   -0.2801032   1.1257495  -0.2488   0.803504
```

```
SystemD:GroupGroup B    0.4839063    1.1457544    0.4223   0.672772
SystemE:GroupGroup B   -0.6965910    1.1253158   -0.6190   0.535904
Very Poorly|Poorly     -0.3962264    0.5599147   -0.7077   0.479160
Poorly|Well             1.7775387    0.6083210    2.9220   0.003477 **
Well|Very Well          4.5998077    0.7448544    6.1754   6.598e-10 ***
---
Signif. codes:  0 '***' 0.001 '**' 0.01 '*' 0.05 '.' 0.1 ' ' 1
```

As you can see the estimates and the standard errors are the same as the OLR output, the t-values have been quoted as z-values, and a p-value has been calculated for each one.

```
# Simplify model and print the final output
mod2 = polr(Likert.Response ~ System + Group, weights = Response,
        data = data28)
summary(mod2); coeftest(mod2)
```

The next model showed that Group was not significant to the model – t-value of 0.025 or a p-value of 0.980.

```
mod3 = polr(Likert.Response ~ System, weights = Response,
        data = data28)
summary(mod3)
```

```
Re-fitting to get Hessian
```

```
Call:
polr(formula = Likert.Response ~ System, data = data28,
        weights = Response)
```

```
Coefficients:
            Value  Std. Error  t value
SystemB   -0.4279      0.5521  -0.7751
SystemC    4.5076      0.7193   6.2666
SystemD    2.2172      0.6157   3.6012
SystemE    2.3173      0.6094   3.8026
```

```
Intercepts:
                   Value  Std. Error  t value
Very Poorly|Poorly -0.3896      0.4005  -0.9728
Poorly|Well         1.7657      0.4644   3.8020
Well|Very Well      4.5560      0.6282   7.2524
```

```
Residual Deviance: 234.7583
AIC: 248.7583
```

The final model showed that System was significant to the model.

The results are given in the log odds scale as with the binomial GLMs and BRGLMs, so the results will need to be transformed on the exponential scale to get the odds ratio. The interpretation of the results is however slightly different due to the ordinal scale. If we used System C instead of System A, the odds of Very Well versus any of the three other Likert.Responses is exp(4.5076) = 90.70 times greater. Likewise the odds of Very Well or Well Versus Poorly or Very Poorly is also 90.70 times greater for System C instead of System A. The same odds value is due to the assumption of linearity, or response variable steps being equal, that I mentioned earlier.

If we used System B instead of System A, the log odds of Very Well versus any of the other Likert.Responses is -0.428, and the odds ratio is exp(-0.428) = 0.652. In other words if we used System A instead of System B, the odds of Very Well versus any other Likert.Response is (1/0.652) = 1.53 times greater. The same odds would apply for Very Well and Well versus Poorly and Very Poorly, as well as for Very Well, Well, and Poorly versus Very Poorly for using System A instead of System B.

The table showing the log odds, odds ratio, and odds tries to clarify the earlier output; however remember that these values apply equally for all "steps" on the response variable.

Difference		Log Odds	Odds Ratio	Odds	Translation
System A	System B	-0.43	0.65	1.53	Use A instead of B odds are 2 times greater
System A	System C	4.51	90.70	90.70	Use C instead of A odds are 91 times greater
System A	System D	2.22	9.18	9.18	Use D instead of A odds are 9 times greater
System A	System E	2.32	10.15	10.15	Use E instead of A odds are 10 times greater
System B	System C	4.94	139.14	139.14	Use C instead of B odds are 139 times greater
System B	System D	2.65	14.08	14.08	Use D instead of B odds are 14 times greater
System B	System E	2.75	15.57	15.57	Use E instead of B odds are 15 times greater
System C	System D	-2.29	0.10	9.88	Use C instead of D odds are 10 times greater
System C	System E	-2.19	0.11	8.94	Use C instead of E odds are 9 times greater
System D	System E	0.10	1.11	1.11	Using E instead of D odds are roughly equal

```
# Calculate confidence intervals
confint(mod3)

              2.5 %      97.5 %
SystemB   -1.523818   0.6516329
SystemC    3.155909   5.9890214
SystemD    1.042077   3.4667028
SystemE    1.155180   3.5549860
```

The confidence intervals around the differences are also calculated on the log-likelihood scale, so if we wanted confidence intervals on these values, we would have to transform them using the exponential. Looking at the confidence intervals around the difference between System A and System B, and also System A and System C for some examples would give us odds ratios of 0.22 (or the inverse which is 4.59) to 1.92, and 23.47 to 399.02, respectively.

The confidence interval for System A and C is easier to interpret—the odds of Very Well versus the other Likert.Responses with System C is 23 times to 399 times greater than with System A. However for the confidence interval for the difference between System A and System B, it's a bit trickier, the odds of Very Well versus the other Likert. Responses with System A is 5 times greater than with System B to System B being 2 times greater than System A. The latter result is highlighted by the Likert plot that suggests no significant difference between System A and System B.

```
# Obtain p-values
coeftest(mod3)
```

Re-fitting to get Hessian

z test of coefficients:

	Estimate	Std. Error	z value	Pr(>|z|)	
SystemB	-0.42790	0.55208	-0.7751	0.4382946	
SystemC	4.50759	0.71931	6.2666	3.691e-10	***
SystemD	2.21721	0.61568	3.6012	0.0003167	***
SystemE	2.31733	0.60940	3.8026	0.0001432	***
Very Poorly|Poorly	-0.38959	0.40050	-0.9728	0.3306686	
Poorly|Well	1.76572	0.46442	3.8020	0.0001435	***
Well|Very Well	4.55596	0.62820	7.2524	4.093e-13	***

Signif. codes: 0 '***' 0.001 '**' 0.01 '*' 0.05 '.' 0.1 ' ' 1

Again this just gives p-values instead of t-values.

```
# Plot the data
library(HH)
plot.likert(System ~ Likert.Response|Group, value = "Response",
        data = data28, layout = c(1,2), as.percent = TRUE,
        main = "How Poorly/Well did the System Perform
        with your Current Equipment?", ReferenceZero = 2.5,
        xlab = expression(bold("Percent")),
        ylab = expression(bold("System")),
        col = c("firebrick3","indianred1","springgreen","forestgreen"))
```

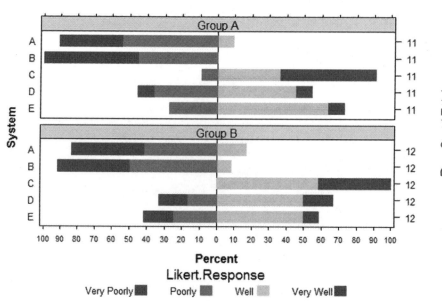

Likert plots are much clearer at displaying ordinal results compared to bar charts and other plots. It highlights that System C clearly displayed the best results, followed by System D and System E, which don't look too dissimilar, followed by System A and System B, which both gave very poor results, but were similar to each other. It also highlights that there isn't much of a difference between the two Groups, so the Likert plot could have been created without splitting into the two Groups, A and B.

In this example the *polr( )* stands for the OLR model, or the proportional odds logistic regression model. In this example I added "*weights = "Freq"*" to the model, this is because I didn't have one row per participant in the dataset I just had the count of responses at each "Likert.Response."

The rest follows the same processes as previous GLM examples with the only differences being that the results are given on the log odds scale and also *t*-values are shown instead of p-values; although the latter can be overcome if required.

Finally a Likert plot was drawn to show its utility at displaying ordinal data along with highlighted differences between "Systems."

The key thing to remember when fitting OLR models is to verify that the ordinal response variable is in the correct order otherwise the model output will be meaningless.

Linear Mixed-Effects Models

Linear mixed-effects models (LMMs) can be thought of as extensions to LMs or GLMs, but instead of fitting only fixed effects as with all previous examples, they can also fit random effects. By using a random effect the assumption of independence is satisfied as a different baseline is assumed for each subject.

LMMs can sometimes be referred to as repeated measures models, as they are generally used when you are repeating measures on a subject. For example the response measurement is taken from the subject at multiple time points, or you have each subject using each piece of equipment, and so forth. Clearly this violates an assumption of the LM in that the responses are not independent, so this needs to be accounted for when fitting a model, hence the random effects.

Another important reason for adding in random effects are that although you may not be interested in the differences between subjects, you need to account for the fact that there are always going to be differences between them—natural variation.

There are two commonly referenced LMMs, those that vary a random effect or more on the intercept and those that vary a random effect or more on the intercept and slope. The first will give different baseline results for let's say all participants. Whereas the second will give different baseline results for all the participants as well as different results due to the explanatory variable chosen, such as exercise as you wouldn't expect everyone to react the same to exercise.

When dealing with a categorical explanatory variable it is difficult to picture a slope. So although this is the official terminology, from now on it will be called the pattern—as it's the change in nonlinear pattern across the variable of interest, time for example.

Nesting is another complication that arises in data sets and this is best dealt with using mixed-effects models, however it's worth noting that it also can be done using LMs.

Nesting occurs when variables sit within other variables, for example if you had taken measurements across six batches (batch) and had taken three measurements from within each batch (repeat), then repeat would be nested within batch. There could also be another layer of nesting added in this example of factory, so repeat would be nested within batch which would be nested within factory.

In this section we will look at three examples, a simple LMM with one random effect on the intercept, a LMM with one random effect on the intercept and pattern, and a more complicated LMM that also includes nesting.

A mixed-effect model will produce t-values instead of p-values as with OLR. If you are happy with interpreting t-values then the *lme4* package will need to be loaded, however if you would like to compute approximate p-values from the t-values, the *lmerTest* package will need to be loaded in R. The latter package uses the Satterthwaite approximation to calculate degrees of freedom; I will use this latter package for ease of interpretation.

The LMMs can be fit using either restricted maximum likelihood (REML) or maximum likelihood (ML). The best way of describing the difference between the two is the REML estimates for the fixed effect are biased and the estimates for the random effects are unbiased, whereas the ML estimates for the fixed effects are unbiased and the estimates for the random effects are biased. In the examples I have used the ML as it gives the AIC, BIC, log likelihood, deviance, and residual degrees of freedom whereas the REML only gives the REML criterion.

Example 7.10 investigates the heart rate "Heart.Rate" of subjects "Subject" for two different treatments "Treatment." Each subject completed the course twice per treatment and their heart rate was recording from resting, 0 seconds, and then every 10 seconds until the end of the experiment at 60 seconds. There was a prolonged period of rest between each completion of the course. This example looks at whether the time or treatment had an effect on subject's heart rate. It will fit a varying intercept model using the "Subject" random effect variable as each "Subject" will have a different baseline "Heart.Rate."

EXAMPLE 7.10

We have a response variable of Heart.Rate, two explanatory variables of Treatment and Time, and a random effect of Subject. The response variable is continuous, and the explanatory variables and random effect are discrete. We are looking to see if Treatment, Time, or the interaction between the two has a relationship with the Heart. Rate of Subjects.

```
# Input the data
Heart.Rate = c(66,80,92,94,124,159,181,66,79,87,92,144,147,168,67,80,90,
               97,124,161,190,66,80,89,94,143,148,168,54,70,104,128,165,
               174,189,52,64,91,110,120,164,194,55,68,108,126,166,176,189,
               50,66,91,115,121,165,195,54,78,89,121,144,172,200,53,76,83,
               109,154,165,179,53,80,89,125,145,170,198,51,78,83,112,151,
               167,180,70,88,95,100,121,168,175,72,80,88,110,148,171,178,
               71,90,94,104,125,170,175,70,81,90,110,147,172,180,49,87,91,
               115,131,158,178,51,61,64,103,146,167,168,50,89,90,117,132,
               159,180,48,59,64,108,145,169,171,64,87,94,115,135,160,196,
               68,81,97,108,138,155,193,66,87,92,117,135,162,194,67,80,95,
               108,140,157,195,66,87,95,130,158,171,193,63,67,96,110,148,
               154,194,65,86,97,129,160,171,195,63,66,100,110,146,156,195,
               55,86,107,116,132,172,180,55,68,80,100,107,155,182,57,88,
               106,118,135,176,182,55,70,81,100,105,160,183,65,68,88,110,
               139,182,187,66,74,82,90,108,155,189,66,70,86,107,138,185,
               189,67,75,86,91,110,157,189,53,67,94,121,141,183,199,52,64,
               75,103,111,158,169,51,68,97,125,141,185,200,53,66,73,101,
               113,159,175)
Subject = rep(c(1:10), each = 28)
Treatment = rep(c("A","B"), each = 7, 20)
Time = rep(c(0,10,20,30,40,50,60), 40)
data29 = data.frame(Heart.Rate, Subject, Treatment, Time)

# Make sure variables are discrete
data29$Time = factor(data29$Time)
data29$Subject = factor(data29$Subject)

# Fit the full mixed effects model and print the anova output
library(lmerTest)
mod = lmer(Heart.Rate ~ Treatment*Time + (1|Subject), REML = F,
           data = data29)
anova(mod)
```

```
Analysis of Variance Table of type III  with  Satterthwaite
approximation for degrees of freedom
                Sum Sq  Mean Sq  NumDF  DenDF  F.value    Pr(>F)
Treatment         4290     4290      1    270    52.74  4.076e-12 ***
Time            525357    87560      6    270  1076.43  < 2.2e-16 ***
Treatment:Time     848      141      6    270     1.74     0.1124
---
Signif. codes:  0 '***' 0.001 '**' 0.01 '*' 0.05 '.' 0.1 ' ' 1
```

The ANOVA table showed that the interaction between Treatment and Time was not significant to the model, p-value of 0.112 and therefore could be removed from the model.

```
# Simplify model and print the anova output and the summary output
mod2 = lmer(Heart.Rate ~ Treatment + Time + (1|Subject),
       REML = F, data = data29)
anova(mod2)
```

Analysis of Variance Table of type III with Satterthwaite approximation for degrees of freedom

	Sum Sq	Mean Sq	NumDF	DenDF	F.value	Pr(>F)	
Treatment	4290	4290	1	270	50.78	9.422e-12	***
Time	525357	87560	6	270	1036.41	< 2.2e-16	***

Signif. codes: 0 '***' 0.001 '**' 0.01 '*' 0.05 '.' 0.1 ' ' 1

When the model was refitted it showed that both Treatment and Time were significant to the model at the 99% confidence level.

```
summary(mod2)
```

Linear mixed model fit by maximum likelihood t-tests use Satterthwaite
  approximations to degrees of freedom [lmerMod]
Formula: Heart.Rate ~ Treatment + Time + (1 | Subject)
  Data: data29

AIC	BIC	logLik	deviance	df.resid
2072.1	2108.4	-1026.0	2052.1	270

Scaled residuals:

Min	1Q	Median	3Q	Max
-2.70142	-0.69578	-0.01159	0.68426	2.60040

Random effects:

Groups	Name	Variance	Std.Dev.
Subject	(Intercept)	10.81	3.289
Residual		84.48	9.191

Number of obs: 280, groups: Subject, 10

Fixed effects:

	Estimate	Std. Error	df	t value	Pr(>|t|)	
(Intercept)	63.539	1.870	58.780	33.986	< 2e-16	***
TreatmentB	-7.829	1.099	270.000	-7.126	9.42e-12	***
Time10	16.350	2.055	270.000	7.955	4.95e-14	***
Time20	30.200	2.055	270.000	14.694	<2e-16	***
Time30	50.350	2.055	270.000	24.498	< 2e-16	***
Time40	76.275	2.055	270.000	37.112	< 2e-16	***
Time50	105.750	2.055	270.000	51.453	< 2e-16	***
Time60	125.750	2.055	270.000	61.184	< 2e-16	***

```
Correlation of Fixed Effects:
            (Intr)  TrtmnB  Time10  Time20  Time30  Time40  Time50
TreatmentB  -0.294
Time10      -0.550  0.000
Time20      -0.550  0.000   0.500
Time30      -0.550  0.000   0.500   0.500
Time40      -0.550  0.000   0.500   0.500   0.500
Time50      -0.550  0.000   0.500   0.500   0.500   0.500
Time60      -0.550  0.000   0.500   0.500   0.500   0.500   0.500
```

The output of the random effects shows that the variability in the Heart.Rate due to Subjects was 3.29 standard deviations, and that of unknown variables is 9.19 standard deviations. This could be due to something like the different pieces of recording equipment, different temperature, or weather conditions, and so forth, which was not recorded.

The output of the fixed effects shows that the average resting Heart.Rate, Time 0, for Treatment-A was 63.54 and that of Treatment-B was (63.539 – 7.829) = 55.71. For examples of other calculations the average Heart.Rate with Treatment-A at 30 seconds was (63.539 + 50.350) = 113.89, and the average Heart.Rate with Treatment-B at 60 seconds was (63.539 – 7.829 + 125.75) = 181.46.

```
# Calculate confidence intervals
confint(mod2)
```

```
                  2.5 %       97.5 %
.sig01         1.840527     5.962966
.sigma         8.467535    10.024775
(Intercept)   59.807072    67.271499
TreatmentB    -9.989453    -5.667690
Time10        12.307362    20.392638
Time20        26.157362    34.242638
Time30        46.307362    54.392638
Time40        72.232362    80.317638
Time50       101.707362   109.792638
Time60       121.707362   129.792638
```

The confidence intervals show the 95% intervals around the estimates, with .sig01 referring to the standard deviation of the Subject random effect, 1.84 to 5.96 standard deviations, and .sigma referring to the standard deviation of the residual random effect, 8.47 to 10.02 standard deviations.

```
# Check assumptions
qqnorm(resid(mod2)); qqline(resid(mod2))
plot(mod2)
```

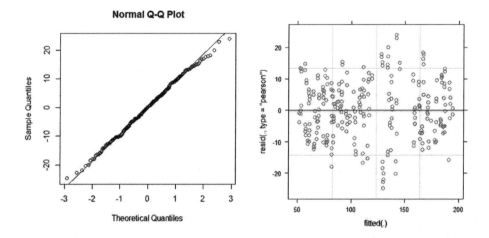

The diagnostic plots show that the residuals are normally distributed and that the variability in the errors was consistent.

```
# Print baseline values for each subject
coef(mod2)
```

```
$Subject
   (Intercept) TreatmentB  Time10  Time20  Time30  Time40  Time50  Time60
1    60.40626  -7.828571   16.35    30.2   50.35  76.275  105.75  125.75
2    65.82344  -7.828571   16.35    30.2   50.35  76.275  105.75  125.75
3    65.51628  -7.828571   16.35    30.2   50.35  76.275  105.75  125.75
4    65.06950  -7.828571   16.35    30.2   50.35  76.275  105.75  125.75
5    59.68024  -7.828571   16.35    30.2   50.35  76.275  105.75  125.75
6    66.27022  -7.828571   16.35    30.2   50.35  76.275  105.75  125.75
7    68.64372  -7.828571   16.35    30.2   50.35  76.275  105.75  125.75
8    61.38358  -7.828571   16.35    30.2   50.35  76.275  105.75  125.75
9    61.60697  -7.828571   16.35    30.2   50.35  76.275  105.75  125.75
10   60.99265  -7.828571   16.35    30.2   50.35  76.275  105.75  125.75

attr(,"class")
[1] "coef.mer"
```

The average Heart.Rates we calculated using the coefficients assume a constant baseline across Subjects that we have actually amended by including Subject as a random effect. So looking at the coefficients output we can see the amended calculations per subject. Here you can see the intercept, resting Heart.Rate for Treatment A, changes per Subject, however the Treatment and Time coefficients remain the same. We expect the Treatment effect to be constant across Subjects, but probably the Time effect will vary per subject, which is also apparent in the plot of the data, in which case we need to refit the model to allow for random pattern varying intercept and slope/pattern LMM.

```
# Plot the data
ggplot(data29, aes(x = Time, y = Heart.Rate, colour = Treatment,
        shape = Treatment)) + theme_bw() +
        facet_wrap(~Subject) + geom_point()
```

Heart Rate by Time and Treatment per Subject

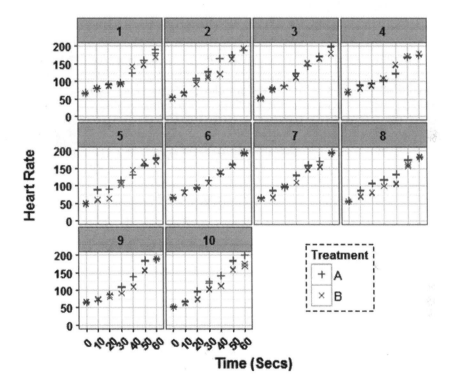

The scatterplot shows the Heart.Rate per Subject by Time and Treatment. It's clear to see the upward trend of Heart.Rate over Time, which is to be expected, and also shows some differences between the two Treatments.

In this example the *lmer()* stands for the LMM. The response is entered first followed by the fixed effects and any interactions. Then the random effects are entered, here the random effect is *(I | Subject)* and this means that only the intercept will vary per "Subject." I also specified that *"REML = F"*, which means that the model will give me the ML estimates. Finally the data set used is specified.

The rest follows the same processes as previous GLM examples with the only differences being that the results shown will be for both the random effects and the fixed effects.

Coefficients are calculated for each "Subject" but you can see that the "Treatment" and the "Time" remains constant for each "Subject."

Finally a scatter plot was drawn to show the trend of "Heart.Rate" over "Time" for each "Subject" and "Treatment."

The key thing to remember when fitting LMMs is to consider what your random effects are; if they are appropriate; and if just a varying intercept, baseline, is adequate for your question.

In Example 7.10 it is probably unrealistic to assume that each subject's heart rate will not vary given time, as everyone's body works differently; so Example 7.11 will look at including this random pattern due to time. This example will fit a varying pattern and varying intercept model using the "Subject" variable, varying the pattern of the "Subject" variable by "Time." Each subject will have a different baseline heart rate and a different heart rate at each time point.

EXAMPLE 7.11

We will be using the same data as in the last example, but this time we will allow the pattern of the Time variable to vary by Subject. We are still looking to see if Treatment or Time has a relationship with the Heart.Rate of Subjects.

```
# Fit the final mixed effects model and print the anova output
# and the summary output
mod3 = lmer(Heart.Rate ~ Treatment + Time + (1 + Time|Subject),
        REML = F, data = data29)
anova(mod3)
```

Analysis of Variance Table of type III with Satterthwaite approximation for degrees of freedom

	Sum Sq	Mean Sq	NumDF	DenDF	F.value	Pr(>F)	
Treatment	4290	4290	1	230.00	79.58	<2.2e-16	***
Time	238946	39824	6	12.95	738.74	1.221e-15	***

Signif. codes: 0 '***' 0.001 '**' 0.01 '*' 0.05 '.' 0.1 ' ' 1

The ANOVA table showed that both Treatment and Time were significant to the model at the 99% confidence level.

```
summary(mod3)
```
Linear mixed model fit by maximum likelihood t-tests use Satterthwaite
  approximations to degrees of freedom [lmerMod]
Formula: Heart.Rate ~ Treatment + Time + (1 + Time | Subject)
   Data: data29

AIC	BIC	logLik	deviance	df.resid
2052.2	2186.7	-989.1	1978.2	243

```
Scaled residuals:
     Min         1Q     Median        3Q       Max
-2.43272   -0.68328    0.00068   0.64156   2.76617
```

Random effects:

Groups	Name	Variance	Std.Dev.	Corr					
Subject	(Intercept)	43.60	6.603						
	Time10	14.72	3.837	-0.69					
	Time20	37.46	6.120	-0.69	0.38				
	Time30	147.86	12.160	-0.86	0.60	0.90			
	Time40	141.41	11.892	-0.62	0.65	0.66	0.84		
	Time50	90.23	9.499	-0.93	0.47	0.69	0.84	0.50	
	Time60	79.40	8.910	-0.72	0.26	0.94	0.92	0.70	0.75
Residual		53.91	7.342						

Number of obs: 280, groups: Subject, 10

Fixed effects:

	Estimate	Std. Error	df	t value	Pr(>|t|)	
(Intercept)	63.5393	2.4291	10.9300	26.157	3.32e-11	***
TreatmentB	-7.8286	0.8776	230.0000	-8.921	2.22e-16	***
Time10	16.3500	2.0415	13.4100	8.009	1.80e-06	***
Time20	30.2000	2.5379	10.9200	11.899	1.36e-07	***
Time30	50.3500	4.1811	10.0000	12.042	2.82e-07	***
Time40	76.2750	4.1032	10.1000	18.589	3.86e-09	***
Time50	105.7500	3.4232	10.2800	30.893	1.78e-11	***
Time60	125.7500	3.2611	10.3100	38.560	1.74e-12	***

Correlation of Fixed Effects:

	(Intr)	TrtmnB	Time10	Time20	Time30	Time40	Time50
TreatmentB	-0.181						
Time10	-0.622	0.000					
Time20	-0.668	0.000	0.434				
Time30	-0.815	0.000	0.483	0.759			
Time40	-0.621	0.000	0.517	0.590	0.789		
Time50	-0.861	0.000	0.437	0.615	0.774	0.495	
Time60	-0.705	0.000	0.337	0.783	0.828	0.657	0.686

The LMM output of the random effects shows that the variability in the Heart.Rate due to Subjects was 6.60 standard deviations, and that of unknown variables is 7.34 standard deviations. There are also contributions to the random effects from the different Time points.

The output of the fixed effects shows that the average resting Heart.Rate for Treatment-A was 63.54 and that of Treatment-B was $(63.5393 - 7.8286) = 55.71$. For examples of other calculations, the average Heart.Rate with Treatment-A at 30 seconds was $(63.5393 + 50.35) = 113.89$, and the average Heart.Rate with Treatment-B at 60 seconds was $(63.5393 - 7.8286 + 125.75) = 181.46$. This output is the same as the varying intercept LMM in Example 7.10 as they are average estimates, however the coefficients per Subject will be different.

```
# Check assumptions
qqnorm(resid(mod3)); qqline(resid(mod3))
plot(mod3)
```

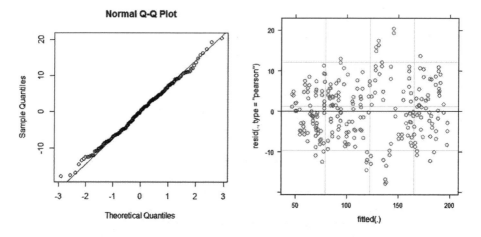

The diagnostics plots show that the residuals are normally distributed and that the variability in the errors was consistent.

```
# Print baseline values for each subject
coef(mod3)
```

```
$Subject
   (Intercept) TreatmentB  Time10  Time20  Time30  Time40   Time50   Time60
1       70.653     -7.829  15.581  20.640  31.550  65.406   90.933  111.555
2       57.853     -7.829  17.101  39.490  66.519  87.837  115.137  139.418
3       58.794     -7.829  19.687  33.923  62.561  92.012  111.608  132.156
4       71.409     -7.829  14.972  23.062  37.331  67.105   97.483  113.708
5       54.748     -7.829  22.983  29.621  58.667  87.020  114.054  125.620
6       70.108     -7.829  13.722  30.112  43.849  72.922   95.700  124.191
7       66.121     -7.829  15.927  34.119  56.475  89.041  101.726  131.783
8       62.390     -7.829  16.385  31.689  48.008  64.996  108.233  124.466
9       67.790     -7.829  10.577  25.133  39.082  60.524  103.971  121.440
10      55.946     -7.829  16.563  34.211  59.458  75.887  118.655  133.165

attr(,"class")
[1] "coef.mer"
```

The average Heart.Rates we calculated assume a constant baseline across Subjects that we have amended by including Subject as a random effect. So looking at the coefficients output we can see the amended calculations per subject. Here you can see the intercept, resting Heart.Rate for Treatment A, changes per Subject, along with the pattern at each Time measurement. However the Treatment coefficients remain the same. So using the examples given earlier but for a specific Subject, say Subject 1, the resting Heart.Rate with Treatment A was 70.65 and for Treatment B was $(70.65331 - 7.828571) = 62.82$. Treatment A at 30 seconds was $(70.65331 +$

31.55024) = 102.20, and Treatment B at 60 seconds was (70.65331 − 7.828571 + 111.5545) = 174.38. If you compare the same four readings (70.65, 62.82, 102.20, and 174.38) to the same readings for Subject 10 (55.95, 48.12, 115.40, and 181.28), you can see the differences.

This example followed exactly the same process as Example 7.10 except that instead of just varying the intercept by "Subject," it also varied the pattern of "Time" by "Subject."

Again, the key thing to remember when fitting LMMs is to consider what your random effects are; if they are appropriate; and if just a varying intercept, baseline, is adequate for your question. Was it appropriate that subjects had different starting heart rates? Yes. Was it more appropriate that the heart rate per time point went up consistently across subjects or that the heart rate was allowed to vary per subject? The latter.

Example 7.12 looks at a LMM that includes nesting. The test scores "Test. Score" for different school subjects, English and Maths, "Subjects" have been recorded to see if there are any significant differences between the results. The test scores were recorded for a number of students from two different classes "Class" within three different schools "School." Each class contained twenty students that recorded their English and Maths test scores.

This example will fit two models; a varying intercept model as well as a varying pattern and varying intercept model, both using the nested "Class" and "School" variables. The second LMM varying the slope of the nested variables by "Subject."

EXAMPLE 7.12

We have a response variable of Test.Score, an explanatory variable of Subject, and two random effects of School and Class, with Class nested within School. The response variable is continuous, and the explanatory variable and random effects are discrete. We are looking to see if the different Subjects result in different Test.Scores.

```
# Input the data
Test.Score = c(94,88,86,90,94,87,87,92,89,92,87,94,93,91,89,92,91,91,95,91,
               82,84,90,81,92,89,85,94,88,94,94,94,86,94,93,84,82,92,92,
               83,89,83,81,87,84,80,81,83,88,82,81,90,82,85,87,82,86,84,
               87,88,82,91,95,77,88,87,79,75,91,77,82,91,95,92,89,83,79,
               90,83,83,82,79,79,78,83,82,81,77,80,79,84,83,81,78,77,75,
               76,76,84,75,78,78,71,79,70,75,75,78,76,71,76,76,73,71,80,
               70,71,78,71,74,76,74,74,77,81,78,79,76,82,79,80,73,72,83,
               72,81,81,72,79,74,67,75,71,66,65,71,73,69,65,67,71,72,68,
               73,65,65,74,67,72,72,82,70,72,86,89,87,87,88,74,92,70,89,
               86,63,68,74,88,71,88,91,76,86,75,79,76,69,86,71,78,67,67,
```

```
                73,69,81,79,78,80,72,81,69,72,75,76,68,72,78,78,77,71,73,
                70,77,75,75,69,77,74,76,68,78,76,75,68,74,69,78,76,70,79,
                78,67,65,86,88,65,88,73, 66,65,85)
School = rep(c("A","B","C"), each = 80)
Class = rep(c("1","2"), each = 20,6)
Subject = rep(c("English","Maths"), each = 40, 3)
data30 = data.frame(Test.Score, School, Class, Subject)

# Make sure variables are discrete
data30$Class = factor(data30$Class)

# Fit the nested mixed effects model varying the intercept and
# print the summary output
mod = lmer(Test.Score ~ Subject + (1|School/Class), REML = F,
       data = data30)
summary(mod)
```

```
Linear mixed model fit by maximum likelihood t-tests use
Satterthwaite
  approximations to degrees of freedom [lmerMod]
Formula: Test.Score ~ Subject + (1 | School/Class)
   Data: data30

   AIC    BIC  logLik deviance df.resid
 1500.0  1517.4  -745.0   1490.0     235

Scaled residuals:
    Min     1Q  Median     3Q     Max
 -3.1638 -0.6607  0.1033  0.5817  2.8737

Random effects:
 Groups          Name  Variance Std.Dev.
 Class:School (Intercept)   6.909    2.629
 School       (Intercept)  26.500    5.148
 Residual                  26.676    5.165
Number of obs: 240, groups:  Class:School, 6;  School, 3

Fixed effects:
             Estimate Std. Error      df t value Pr(>|t|)
(Intercept)   81.6083    3.1949   3.0700  25.543 0.000113 ***
SubjectMaths  -4.1083    0.6668 234.0000  -6.161 3.12e-09 ***
---
Signif. codes:  0 '***' 0.001 '**' 0.01 '*' 0.05 '.' 0.1 ' ' 1

Correlation of Fixed Effects:
            (Intr)
SubjectMths -0.104
```

The output of the varying intercept model showed that Subject was significant to the model at the 99% confidence level. The output of the random effects showed that the variability in the Test.Score due to School was 5.15 standard deviations, the variability due to Class within School was 2.63 standard deviations, and the variability due to unknown variables was 5.17 standard deviations.

The output of the fixed effects shows that the average Test.Score for English was 81.61 and that of Maths was (81.6083 − 4.1083) = 77.50.

```
# Calculate confidence intervals
confint(mod)
```

	2.5 %	97.5 %
.sig01	1.210701	8.058294
.sig02	0.000000	16.136540
.sigma	4.730076	5.670659
(Intercept)	72.723753	90.492950
SubjectMaths	-5.420591	-2.796076

The confidence intervals show the 95% intervals around the estimates, with .sig01 referring to the standard deviation of the Class within School random effect, 1.21 to 8.06 standard deviations; .sig02 referring to the standard deviation of the School random effect, 0.00 to 16.14 standard deviations; and .sigma referring to the standard deviation of the residual random effect, 4.73 to 5.67 standard deviations.

```
# Check assumptions
qqnorm(resid(mod)); qqline(resid(mod))
plot(mod)
```

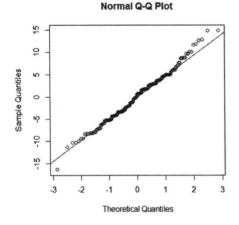

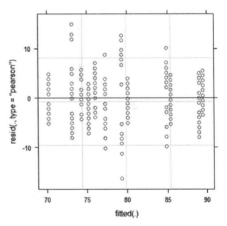

The assumptions plots show that the residuals are normally distributed and that the variability in the errors was consistent.

```
# Calculate predicted values for each subject
pred = with(data30,expand.grid(Class = levels(Class),
        Subject = levels(Subject), School = levels(School)))
pred$Test.Score = predict(mod, newdata = pred)
pred
```

	Class	Subject	School	Test.Score
1	1	English	A	89.52065

2	2	English	A	89.04186
3	1	Maths	A	85.41231
4	2	Maths	A	84.93353
5	1	English	B	80.12441
6	2	English	B	74.35618
7	1	Maths	B	76.01607
8	2	Maths	B	70.24784
9	1	English	C	79.34083
10	2	English	C	77.26609
11	1	Maths	C	75.23249
12	2	Maths	C	73.15775

To obtain average Test.Scores using the model we use the **predict()** command. So looking at the results we can get a rough idea about the average Test.Scores within School and Class. At School A we can see the overall average English Test. Score was ((89.52+89.04)/2) = 89.28 and the overall average Maths Test.Score was ((85.41+84.93)/2) = 85.17. We can take averages of averages as there were equal sample sizes within each group. We shouldn't take averages of the Class ignoring School as Class is nested within School.

```
# Fit the nested mixed effects model varying the slope and the
intercept
# and print  the summary output
mod2 = lmer(Test.Score ~ Subject + (1 + Subject|School/Class),
       REML = F, data = data30)
summary(mod2)
```

```
Linear mixed model fit by maximum likelihood t-tests use
Satterthwaite
  approximations to degrees of freedom [lmerMod]
Formula: Test.Score ~ Subject + (1 + Subject | School/Class)
   Data: data30

   AIC    BIC  logLik deviance df.resid
1507.0 1538.3  -744.5   1489.0      231

Scaled residuals:
    Min      1Q  Median      3Q     Max
-3.2745 -0.6834  0.0497  0.6448  2.7733

Random effects:
 Groups                   Name Variance Std.Dev.  Corr
 Class:School      (Intercept)  7.2247   2.6879
                  SubjectMaths  1.5559   1.2474  -0.23
 School            (Intercept) 28.5719   5.3453
                  SubjectMaths  0.1446   0.3803  -1.00
 Residual                      26.2401   5.1225
Number of obs: 240, groups:  Class:School, 6;  School, 3
```

```
Fixed effects:
               Estimate  Std. Error      df  t value  Pr(>|t|)
(Intercept)     81.6083      3.3086  3.0030    24.67  0.000145 ***
SubjectMaths    -4.1083      0.8631  5.3320    -4.76  0.004277 ***
---
Signif. codes:  0 '***' 0.001 '**' 0.01 '*' 0.05 '.' 0.1 ' ' 1

Correlation of Fixed Effects:
            (Intr)
SubjectMths -0.359
```

Repeating the model but this time varying the Subject pattern by School and Class, as the Subject Test.Scores will probably vary across the Schools and Classes due to teaching methods, still showed that Subject was significant to the model at the 99% confidence level.

The output of the random effects showed that the variability in the Test.Score due to School was 5.35 standard deviations, that due to Class within School was 2.69 standard deviations, and that due to unknown variables was 5.12 standard deviations. It also shows the contribution of the Subject to the random effects.

The output of the fixed effects shows that the average Test.Score for English was 81.61 and that of Maths was (81.6083 − 4.1083) = 77.50, which is the same as the last model.

```
# Check assumptions
qqnorm(resid(mod2)); qqline(resid(mod2))
plot(mod2)
```

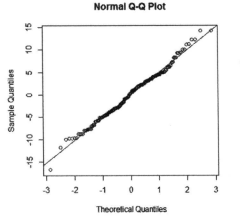
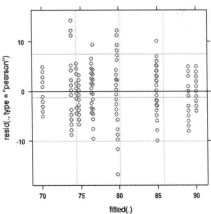

The assumptions plots show that the residuals are normally distributed and that the variability in the errors was consistent.

```
# Calculate predicted values for each subject
pred2 = with(data30,expand.grid(Class = levels(Class),
```

```
        Subject = levels(Subject), School = levels(School)))
pred2$Test.Score = predict(mod2, newdata = pred2)
pred2
```

```
   Class  Subject  School  Test.Score
1    1    English     A     90.05978
2    2    English     A     89.03245
3    1     Maths      A     84.87723
4    2     Maths      A     84.94755
5    1    English     B     79.69718
6    2    English     B     74.45768
7    1     Maths      B     76.44381
8    2     Maths      B     70.13846
9    1    English     C     79.77354
10   2    English     C     76.62937
11   1     Maths      C     74.79930
12   2     Maths      C     73.79365
```

To obtain average Test.Scores using the model we use the **predict()** command. So looking at the results we can get a rough idea about the average Test.Scores within School and Class. At School A we can see the overall average English Test. Score was ((90.06+89.03)/2) = 89.55 and the overall average Maths Test.Score was ((84.88+84.95)/2) = 84.91. We can take averages of averages as there were equal sample sizes within each group. We shouldn't take averages of the Class ignoring School as Class is nested within School. It's also worth noting that these results are very similar to those produced from the first model, so it would be more preferable to use the simpler model; therefore this would suggest that the varying intercept LMM was adequate.

```
# Plot the data
ggplot(data30, aes(x = Subject, y = Test.Score)) + theme_bw() +
         facet_wrap(~School + Class) +
         stat_boxplot(geom = "errorbar") + geom_boxplot()
```

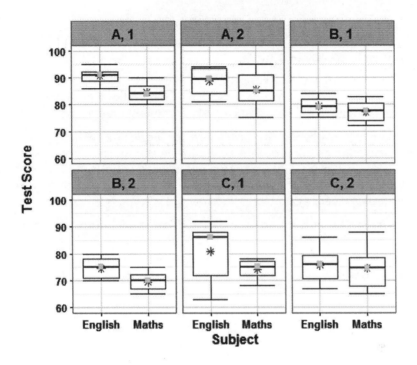

The box plot shows the Test Scores by Subject, School, and Class. It highlights how the Maths scores are generally lower than the English scores, and that there seem to be slight differences between the Schools, but not so much between the Classes within the Schools.

This example followed exactly the same process as Example 7.10 and Example 7.11 with the only difference being the nested effects of "Class" within "School."

Remember to account for mixed effects, random effects, and nested variables.

Summary

The beginning of the chapter described what models are and how they are the next step up from simple hypothesis testing. Models can account for more explanatory variables or more levels within explanatory variables and their relationship to the single response variable. The chapter was then split into two main sections, statistical model components and statistical models.

The statistical model components discussed the elements that make up modeling; it was divided into four sections; model assumptions, model structure, model process, and model output.

Model assumptions looked at the main assumptions for linear models showing not only examples of where the assumptions have been satisfied but also where they have been violated.

The model structure showed the simplest form of linear models and how they can be built to include more explanatory variables, interactions, or polynomial terms. It also introduced the idea of mixed-effect models and nesting.

The model process explained the procedure for beginning with the most complicated model, then simplifying it by removing the nonsignificant terms with the highest p-value and the most complicated interaction. It also included examples of how to do this and presented multiple-comparison testing by using least-squares means contrasts.

The model output section showed examples of linear model output and how to interpret the various sections of output such as the coefficients, the p-values, and the R^2 value. It then highlighted the diagnostic plots for checking assumptions, along with an example of multiple comparison testing, then finally a suitable plot to emphasize the results visually.

The statistical models section described some of the more commonly used models and was split into five main parts, each one describing the model or models, when they should be used, and examples to show how they are used along with translation.

The first statistical model section was simple models, which included linear models (LMs) and analysis of variance (ANOVA). It discussed the use of both with a continuous response variable with both continuous and discrete explanatory variables. It highlighted some of the similarities and differences between the two calculations and ended up recommending using LMs except when a complicated output table needs tidying up for interpretation.

The next section was generalized linear models (GLMs), which included Gaussian GLMs, Poisson GLMs, negative binomial GLMs (NBGLMs), binomial GLMs, and bias-reduction binomial-response GLMs (BRGLMs).

Gaussian GLMs were shown to be no different to LMs, but the example also highlighted how to conduct model simplification, multiple comparison testing, which included interpretation of an interaction output, and also testing between different models.

The Poisson GLM and NBGLM were closely linked, as both cases dealt with a discrete response variable, more specifically count data. They both also gave their output on the log scale and so must be back transformed using the exponential to get the values back to the response scale. The main difference between the two GLMs is that the NBGLM is used for over dispersed count data.

The next two GLMs are also closely linked, the binomial GLM and the BRGLM, as they are both concerned with binary response data. This time both GLMs gave their model output on the log odds scale, so they must be back transformed using the exponential to get the odds ratio scale. As such interpretation can be tricky, and where an odds ratio is less than one, it is advised to switch the explanatory variables and report the relationship in the opposite direction by dividing one by the odds ratio for the positive odds ratio. The main difference between the two GLMs is that the BRGLM should be used when there is clear separation of zeros and ones in different groups.

The third statistical models section described the zero-inflated models: the zero-inflated Poisson (ZIP) model and the zero-inflated negative binomial (ZINB) model. It showed that these models were used when there were significantly more zeros in the data than would be expected for both the Poisson GLM and the NBGLM. The zero-inflated models output is split into two sections, that for the count data which is on the log scale, and that for the zero-inflated data which is on the log odds scale. The example showed how to run and interpret these models and output including how to compare the zero-inflated models to their GLM counterpart by using the Vuong non-nested hypothesis test, as well as how to decide between using the ZIP or the ZINB.

The next statistical model explained was the ordinal logistic regression (OLR) model, which is used with ordinal response data. It discussed that although the output uses t-values instead of p-values, there is a way to calculate p-values if needed. It also highlighted that the values are given on the log odds scale and needed to be translated in terms of the ordinal response, which must be ordered correctly or the model will be useless. It also showed that a Likert plot is very good at displaying ordinal data.

The final statistical models described were linear mixed-effects models (LMMs) which can cover both random effects and nested explanatory variables. Two simple LMMs were shown, one with a varying intercept and one with a varying intercept and pattern. The first means that the random effect variable will have a different baseline per level, but all other variables will remain at a constant rate. Whereas the second means that the random effect variable will not only have a different baseline per level, but also different rates of the chosen fixed effect variable. The third LMM used nesting, measurements taken from a variable that sits within another variable, and looked at both a varying intercept and a varying intercept and pattern model including the nested random effect variables.

Chapter 8 progresses on from univariate responses to multivariate responses and looks at some of the more common multivariate techniques available. With many multivariate methods the interpretation is very subjective, so the chapter will only be a guide as to what possibilities there are rather than instruction to use certain methods on certain data types as in this chapter.

Multivariate Analysis

What Have I Found in My Larger Data?

Multivariate analysis should be used on data with multiple response variables. I have termed the data as *large* as the data doesn't necessarily have to fall into the category of "big data."

For example, a data set with two nonindependent response variables could be analyzed using a multivariate analysis of variance (MANOVA), which is a multivariate technique. However the data is not "big data."

With most multivariate techniques the interpretation of the analysis is quite subjective. The next sections of this chapter look at a few of the more common methods and show examples of what can be done. However, there are no real rules on which methods should be used instead of others due to data type and so forth as there was in Chapter 7.

The aim of multivariate analysis is to investigate any trends or patterns that may be in the data; most of the time condensing high dimensional data down into smaller dimensions for easier interpretation.

© Victoria Cox 2017
V. Cox, *Translating Statistics to Make Decisions*, DOI 10.1007/978-1-4842-2256-0_8

To delve into the different multivariate techniques would require another book in itself, so I have chosen three frequently used methods that are sufficiently different from each other, as there is a lot of overlap with multivariate methods. The three multivariate methods are MANOVA, principal component analysis (PCA), and Q methodology.

Other methods you may have come across, but this is not an exhaustive list, include the following:

- Factor analysis (FA): similar to PCA but uses different mathematics.

- Correspondence analysis (CA): sometimes referred to as the categorical equivalent to PCA.

- Cluster analysis: looks to reduce data to representative cases rather than features as with PCA.

- Linear discriminant analysis (LDA): similar to PCA but tries to model differences between classes and has an assumption of multivariate normality.

- Artificial neural network (ANN): uses a training data set and a testing data set, can do dimension reduction itself or can be run on components from PCA.

- Spatio-temporal analysis: methods for analyzing data with both a spatial (space) and temporal (time) aspect; many still being developed.

As I have stated, there are many multivariate techniques and this chapter only aims to give you a taste of a few differing methods: MANOVA is closely related to ANOVA in interpretation but with multiple response variables instead of one; PCA concerns reducing multiple factors into a few more manageable dimensions/variables; and Q methodology works to group subjects into groups by consensus and distinguishing viewpoints.

Multivariate Analysis of Variance

MANOVA is used when there are two or more response variables and one of more explanatory variables. This is one of the least subjective multivariate methods as it is more closely related to statistical modeling principals.

The difference between ANOVA and MANOVA is that with ANOVA you are testing the difference between the means of the groups, whereas with MANOVA you are testing the difference between vectors of means of the groups.

I frequently use MANOVA on shot data as clearly the X and Y coordinates data can't be classed as independent to each other. By using MANOVA I can look for differences between groups, such as weapons, ammo, distances, and so forth, as well as investigating the accuracy and precision of the shots.

There are several assumptions associated with running a MANOVA, however the data can violate some of the assumptions and still be analyzed using MANOVA:

- Multivariate normality: the dependent response variables must follow multivariate normality for each level within the explanatory variables.

- Linearity: all relationships between all pairs of response variables should be linear.

- Homogeneity of variance–covariance matrix: this assumption will automatically be satisfied with a balanced design. If the design is unbalanced then Box's M test can be used to test the assumption using a p-value of 0.001 as a cutoff.

- Homogeneity of variance: the variance of the response variables should be roughly equal across the explanatory variables. This can be tested with Levene's test and if the assumption is violated the recommendation is to adjust the MANOVA p-value cutoff to 0.025 or 0.01.

- Independence: the assumption that all the subjects/shots are independent.

- Multicollinearity and singularity: there should be moderate correlation between the response variables. General advice is if the correlation is larger than 0.8/0.9 then consider combining the response variables into one, and when the correlation is very low, below 0.1, consider running separate ANOVAs on them.

- Outliers: MANOVA is sensitive to outliers, and as such they should be investigated using Mahalanobis distance. MANOVA can cope with a few outliers if the sample size is large and the outliers aren't too extreme.

- Sample size: the recommended minimum sample size at each combination of explanatory variables is around 20. However, MANOVA can be robust with smaller sample sizes by using a different test statistic.

There are multiple test statistics that can be used with MANOVA, which is different to ANOVA that only uses the F test to generate p-values, but the recommended test statistic to use is Pillai's trace. Pillai's trace is more robust to assumption violations, small sample sizes, and unequal groups, but therefore will give a more conservative result.

MANOVA output will inform whether there's evidence that the explanatory variables are significant to the response variables in general. Follow-up tests can be done, such as running univariate ANOVA's to see if the explanatory variables had more of an effect on one response variable than another. If these are carried out it's recommended to use a p-value of 0.025, if you're working at 95% confidence, as a cutoff to correct for multiple testing. More multiple comparisons can be run at the MANOVA level if you have multiple levels within an explanatory variable, but again these should undergo some multiple comparison testing corrections.

The easiest way to understand MANOVA is to run through an example, so Example 8.1 looks at the X and Y coordinates of shots "Point.X" and "Point.Y" from multiple weapon systems "Weapon" and distances fired "Dist." To check the assumptions, plot confidence ellipses, and also run multiple comparisons tests the following R packages are required; *mvnormtest*, *biotools*, *car*, *mvoutlier*, *RVAideMemoire*, and *shotGroups*.

The *ggplot2* package is required for the graphs, and as before only the basic code will be shown to create the plots. Refer to Chapter 9 to see how to include all other details.

EXAMPLE 8.1

We have two dependent response variables of Point.X and Point.Y and two explanatory variables of Weapon and Dist; both the response variables are continuous and both the explanatory variables are discrete. We are looking to see if Weapon, Dist, or the interaction between the two has an effect on the shots fired.

```
# Input the data
Point.X = c(-43,-40,-28,16,-30,-29,-44,-36,-32,-31,2,-8,8,12,-5,-6,
            11,-7,11,22,19,5,-1,16,2,15,-17,-7,6,-15,15,-68,2,-80,-26,
            -42,-43,-46,-52,-22,-67,-25,-80,-23,-54,-62,-83,4,-46,-12,
            33,8,8,23,18,15,14,-24,2,-8,15,32,34,15,-14,-7,-6,12,-5,42,
            -19,9,-10,7,45,-8,-31,2,-10,8,10,6,14,-12,33,22,-12,-7,10,
            0,-13,-77,-54,-95,-85,-67,-61,-56,-69,-27,-87,-85,-55,-80,
            8,-13,-75,-120,-77,-13,-50,60,-3,-25,35,56,22,45,-50,35,42,
            9,-13,52,-10,52,35,0,-15,62,25,17,-17,-23,10,11,34,10,35,
            -17,51,5,-31,32,-37,0,-7,-2,11,-62,-103,48,15,-55,-62,20,
            -56,-114,56,-21,-142,10,-65,-76,-87,-105,-168,-66,58,-100,4,
            74,-55,-68,43,-44,63,15,-63,31,95,23,-79,-9,12,15,70,48,-5,
            96,-49,-12,0,75,31,-10,21,34,17,-32,-18,6,-20,20,18,16,60,
            33,4,-29)
Point.Y = c(76,42,25,11,44,27,39,17,38,39,55,28,36,33,37,38,26,45,
            25,37,54,22,36,40,90,41,47,63,82,56,9,52,-30,-20,-30,29,25,
```

```
               33,-48,-2,13,35,47,34,-35,3,26,10,-25,16,-5,15,-20,24,1,
               -12,-14,15,35,-16,15,-44,20,-21,-20,4,2,27,58,-22,-30,-8,0,
               45,-24,-29,19,-53,12,0,25,-44,54,-26,6,-18,-47,-9,-3,-53,
               -43,-6,-66,-39,-14,-33,-6,-46,-91,-37,-53,-56,-65,-101,-24,
               -43,-58,-29,-31,-43,-37,-17,-75,20,-87,-30,-100,-88,25,-55,
               -91,-92,-27,-42,-81,-11,27,-45,-58,-65,-129,16,-48,-98,-4,
               -103,-171,-12,-23,-107,-100,-148,-48,19,-32,-35,-35,-52,-46,
               -142,-210,-176,-73,-81,-195,-157,-120,-116,-112,-58,-95,
               -135,-143,-135,-166,-120,-118,-141,-201,-254,-123,-91,-110,
               -143,-145,-232,-185,-166,-198,-122,-183,-96,-165,-113,-158,
               -120,-193,-220,-104,-93,-239,-171,-74,-180,-169,-127,-201,
               -154,-211,-122,0,3,-101,-143,-43,-158,-250,-136,-131,-130)
Dist = rep(c(5,10,15,20), c(30,60,60,60))
W1 = rep(c("A","B","C"), each = 10); W2 = rep(c("A","B","C"),
         each = 20, 3)
Weapon = c(W1, W2)
data31 = data.frame(Point.X, Point.Y, Weapon, Dist)
data31$Dist = factor(data31$Dist)

# Plot the data
library(ggplot2)
ggplot(data31, aes(x = Point.X, y = Point.Y, colour = Weapon,
        shape = Weapon)) + facet_wrap( ~ Dist, ncol = 2) +
        geom_point() + theme_bw()
ggplot(data31, aes(x = Point.X, y = Point.Y, colour = Dist,
        shape = Dist)) + facet_wrap( ~ Weapon, ncol = 2) +
        geom_point() + theme_bw()
```

X and Y Coordinates by Distance and Weapon

X and Y Coordinates by Weapon and Distance

The initial plots potentially show some differences between Weapon A and the other Weapons, however this seems less clear at a Distance of 20m. In addition there seems to be some differences between the shots across the Distances, with the variation increasing as the Distance increases.

```
# Check the assumptions - only run on first combinations
# Multivariate normality
library(mvnormtest)
mshapiro.test(t(data31[data31$Weapon == "A" & data31$Dist == "5",
    1:2]))
```

Shapiro-Wilk normality test

data: Z
W = 0.66844, p-value = 0.000366

Where the following assumption tests require repetition for all combinations of Weapon and Distance only the first combination has been shown on the output to save space—the combination of Weapon A and Dist 5. The multivariate normality tests show that all combinations of Weapon and Distance can be assumed to be normally distributed except "Weapon A, Dist 5" and "Weapon A, Dist 10."

```
# Homogeneity of Variance-Covariance Matrix
library(biotools)
boxM(data31[,1:2], data31$Weapon)
```

Box's M-test for Homogeneity of Covariance Matrices

data: data[, 1:2]
Chi-Sq (approx.) = 24.943, df = 6, p-value = 0.0003499

```
boxM(data31[,1:2], data31$Dist)
```

Box's M-test for Homogeneity of Covariance Matrices

data: data[, 1:2]
Chi-Sq (approx.) = 92.827, df = 9, p-value = 4.402e-16

As the design was unbalanced, there were less repeats for Distance 5m, Box's M test was run. The output suggests that we cannot assume homogeneity of the variance-covariance matrix for either Weapon or Distance, therefore we should use Pillai's trace when running the MANOVA.

```
# Homogeneity of Variance
library(car)
leveneTest(Point.X ~ Weapon, data = data31)
```

Levene's Test for Homogeneity of Variance (center = median)
        Df F value   Pr(>F)
group   2   7.7502 0.0005678 ***
       207

Signif. codes: 0 '***' 0.001 '**' 0.01 '*' 0.05 '.' 0.1 ' ' 1

```
leveneTest(Point.Y ~ Weapon, data = data31)
```

Levene's Test for Homogeneity of Variance (center = median)
        Df F value   Pr(>F)
group   2   0.5048   0.6044
       207

Levene's test for Weapon shows that we can assume equal variance for the Y Coordinates, but not for the X Coordinates. The same test for Distance shows than we can't assume equal variance for either the X or Y Coordinates. As such the MANOVA p-value cutoff point should be amended to 0.025 or 0.01.

```
# Multicollinearity and Singularity
cor(data31$Point.X[data31$Weapon == "A" & data31$Dist == "5"],
    data31$Point.Y[data31$Weapon == "A" & data31$Dist == "5"],
    method = "spearman")
```

```
[1] -0.5835893
```

The correlation shows that for Weapon A, there is moderate correlation for Distances of 5m and 10m and very low correlation for 15m and 20m. For both Weapon B and Weapon C there is moderate correlation for all Distances.

```
# Outliers
library(mvoutlier)
aq.plot(data31[data31$Weapon == "A" & data31$Dist == "5", 1:2])
```

```
$outliers
    1      2      3      4      5      6      7      8      9     10
 TRUE  FALSE  FALSE   TRUE  FALSE  FALSE  FALSE  FALSE  FALSE  FALSE
```

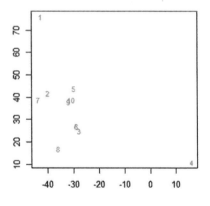

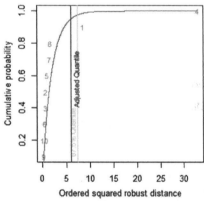

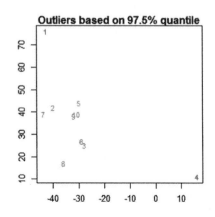

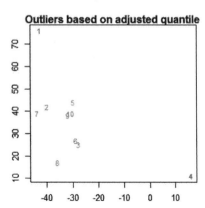

Looking at all the combinations, there are outliers for "Weapon A, Dist 5"; "Weapon A, Dist 10"; "Weapon B, Dist 15"; "Weapon B, Dist 20"; "Weapon C, Dist 15"; and "Weapon C, Dist 20." However, there are only one or two outliers that could be classed as extreme so the decision is to proceed with MANOVA.

```
# Fit the full model and print the output
mod = manova(cbind(Point.X,Point.Y) ~ Dist*Weapon, data31)
summary(mod)
```

```
                Df  Pillai  approx F  num Df  den Df  Pr(>F)
Dist             3  0.73903   38.681       6     396  <2e-16 ***
Weapon           2  0.37189   22.613       4     396  <2e-16 ***
Dist:Weapon      6  0.05997    1.020      12     396  0.4292
Residuals      198
---
Signif. codes:  0 '***' 0.001 '**' 0.01 '*' 0.05 '.' 0.1 ' ' 1
```

The full model output shows that the interaction of Distance and Weapon is not significant to the model, so it can be removed from the model.

```
# Fit the simplified model and print the output
mod2 = manova(cbind(Point.X,Point.Y) ~ Dist + Weapon, data31)
summary(mod2)
```

```
                Df  Pillai  approx F  num Df  den Df       Pr(>F)
Dist             3  0.73335   39.370       6     408  <2.2e-16 ***
Weapon           2  0.36431   22.718       4     408  <2.2e-16 ***
Residuals      204
---
Signif. codes:  0 '***' 0.001 '**' 0.01 '*' 0.05 '.' 0.1 ' ' 1
```

The model output shows that there is evidence that both Distance and Weapon are significant to the model at the 99% confidence level, we were looking for a p-value < 0.025 or 0.01 anyway due to the violation of homogeneity of variance.

```
# Conduct univariate ANOVA's
summary.aov(mod2)
```

```
Response Point.X :
              Df  Sum Sq  Mean Sq  F value  Pr(>F)
Dist           3     954      318    0.252  0.8599
Weapon         2  146519    73260   58.035  <2e-16 ***
Residuals    204  257517     1262
---
Signif. codes:  0 '***' 0.001 '**' 0.01 '*' 0.05 '.' 0.1 ' ' 1
Response Point.Y :
              Df  Sum Sq  Mean Sq  F value  Pr(>F)
Dist           3  906834   302278  183.1591  <2e-16 ***
Weapon         2    1787      893    0.5413  0.5828
Residuals    204  336673     1650
---
Signif. codes:  0 '***' 0.001 '**' 0.01 '*' 0.05 '.' 0.1 ' ' 1
```

The univariate ANOVA shows that there is evidence that Weapon has an effect on the *X* Coordinates at the 99% confidence level, but there's no evidence that Distance does. It also shows that there is evidence that Distance has an effect on the Y Coordinates at the 99% confidence level, but no evidence that Weapon does.

```
# Conduct multiple comparisons testing
library(RVAideMemoire)
pairwise.perm.manova(data31[,1:2], data31$Dist, nperm = 500)
```

```
Pairwise comparisons using permutational MANOVAs (test: Pillai)

data:  data[, 1:2] by data$Dist
500 permutations

        5      10      15
10   0.002      -       -
15   0.002   0.002      -
20   0.002   0.002   0.002

P-value adjustment method: fdr
```

```
pairwise.perm.manova(data31[,1:2], data31$Weapon, nperm = 500)
```

```
Pairwise comparisons using permutational MANOVAs (test: Pillai)

data:  data[, 1:2] by data$Weapon
500 permutations

        A      B
B   0.003      -
C   0.003   0.285

P-value adjustment method: fdr
```

As there are multiple Weapons and multiple Distances, multiple comparisons need to be run to determine where the differences are. All Distances are significantly different to each other at the 99% confidence level. Weapon A is significantly different to both Weapon B and Weapon C at the 99% confidence level, but there is no difference between Weapon B and Weapon C.

```
# Plot the data
ggplot(data31, aes(x = Point.X, y = Point.Y, colour = Weapon,
        shape = Weapon)) + geom_point() + theme_bw()
ggplot(data31, aes(x = Point.X, y = Point.Y, colour = Dist,
        shape = Dist)) + geom_point() + theme_bw()
```

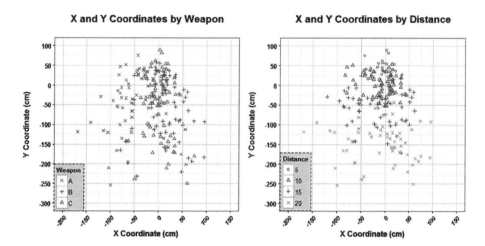

Plotting the data helps to visualize the differences. The plot on the left shows why there was no evidence of a difference on the Y Coordinates, as the spread of all three Weapons is similar, but you can see that for the *X* Coordinates Weapon A shots were more to the left of the aim point (0,0).

The plot on the right emphasizes the differences between the four Distances for the Y Coordinates; it can clearly be seen that as the shooter gets further away from the target, the shots start to drop below the aim point, they also get more dispersed. For the *X* Coordinates, although there was no evidence of a significant difference, you can see that the shots become more dispersed on this axis too.

```
# Add 75% confidence ellipses using the shotGroups shortcut
library(shotGroups)
data31$Series = data31$Weapon
compareGroups(data31, xyTopLeft = FALSE, CEPlevel = 0.75)
data31$Series = data31$Dist
compareGroups(data31, xyTopLeft = FALSE, CEPlevel = 0.75)
```

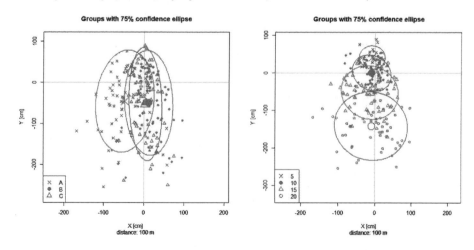

Finally using the shotGroups package to compare the groups will output lots of information and many plots including a box plot of distance to center, the distance to center with Rayleigh mean radius and confidence intervals, the extreme spread bounding boxes, and also confidence ellipses.

The most useful here is the confidence ellipses, these were run at the 95% confidence level and contain 75% of the shots, and both of these values can be amended. Confidence ellipses can be more useful than circular error probable (CEP) and extreme spread as they also show the direction of dispersion. CEP will only draw circles that contain only 50% of the shots, and the extreme spread is very sensitive to outliers and should not be used due to this. For example, in the left plot the data is dispersed along the Y Coordinates more than the *X* Coordinates—this information would have been lost using the CEP. However, the right plot would look very similar to the CEP in terms of shape, but will be larger due to containing 75% of the shots instead of 50%.

Here you can see the use of MANOVA in determining differences between groups for multiple dependent response variables. Some people will just skip the assumptions section and go straight to the MANOVA analysis using Pillai's trace as it is more robust, as long as there is a reasonable sample size, fairly balanced groups, and no extreme outliers, this can be done. However I would recommend that the correlation of the dependent response variables be checked to determine whether MANOVA is needed.

It also shows how useful the confidence ellipses are at highlighting the dispersion of the shots. It emphasizes the direction of the spread, which is an added advantage over other methods such as the CEP, as well as being able to account for whatever percentage of data you are interested in.

Principal Component Analysis

PCA is used when you have a lot of response variables, usually more than the number of responses from subjects. In addition there should be more continuous variables than discrete variables.

The mathematics is very complicated, but basically it tries to convert the raw data into a set of principal components using orthogonal linear transformation. This results in a smaller number of "variables" than the original number of variables, which in turn should be easier to interpret and use in further analysis.

The drawbacks of PCA are that it makes no assumptions about the distribution, the data reduction can result in a loss of information, and it assumes that dimensions with larger variance correspond to interesting features whereas dimensions with smaller variance correspond to "noise."

The first principal component will always account for the largest variation within the data, then the second, then the third, and so forth.

PCA is usually just one stage of analysis, it's used to find the smaller number of components; then they are then taken forward into statistical analysis. This is more of an EDA technique, albeit a very complicated one.

Example 8.2 looks at an example of PCA. This is a simple example to show how the output will look and also some ways of interpreting the results. The data involves students' scores for 10 different school subjects.

There are a few packages in R for running PCA but the one we are using is **FactoMineR**, with **factoextra;** in addition to plot the initial data we are using the **corrplot** package. In more complicated data sets you can add in other variables (discrete or continuous) that you don't want included in the data reduction, but you do want to view visually such as different schools or year groups if we had that extra data in relation to Example 8.2.

EXAMPLE 8.2

The test scores for 24 students across 10 school subjects have been recorded. We are looking to see if there is any relationship between the subjects.

```
# Input the data
English = c(75,71,86,85,40,61,87,94,73,63,59,73,69,82,56,74,55,95,
            87,49,51,72,81,60)
Maths = c(48,90,70,64,70,77,62,68,88,60,76,67,74,50,72,86,76,52,48,
          88,95,31,39,51)
Chemistry = c(50,95,74,63,68,71,78,52,88,90,63,85,82,61,65,84,62,
              53,46,92,82,35,43,53)
Biology = c(60,77,79,67,61,80,71,74,72,99,64,56,87,64,90,79,61,48,
            55,84,84,39,37,54)
Physics = c(52,90,71,61,73,72,54,55,92,57,80,62,99,52,83,76,60,45,
            46,86,92,32,36,52)
French = c(80,70,87,81,53,85,93,96,54,97,64,81,55,62,61,81,75,91,95,
           56,60,76,79,64)
Spanish = c(76,75,85,80,41,57,79,85,89,69,61,64,77,83,69,82,50,96,
            86,48,50,71,82,63)
History = c(81,67,79,90,48,81,99,92,69,51,54,96,62,81,73,40,35,70,
            68,78,71,76,84,94)
Geography = c(55,65,75,85,75,60,63,50,70,78,60,76,67,59,61,45,41,
              55,46,88,92,79,73,95)
Art = c(59,83,76,95,78,66,85,76,59,55,53,65,63,86,54,41,20,51,40,90,
        99,74,87,98)
data32 = data.frame(English, Maths, Chemistry, Biology, Physics,
                    French, Spanish, History, Geography, Art)

# Plot the initial data
library(corrplot)
cor.mat = round(cor(data32),2)
corrplot(cor.mat, type = "lower", order = "FPC", tl.col = "black",
         tl.srt = 45, diag = F, outline = T)
```

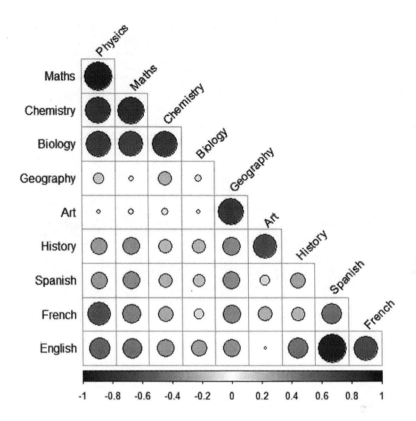

This correlation plot was drawn ordering the School Subjects by the first Principal Component (PC1) scores. We can already see the groupings of Subjects by the size of the circles, the larger the circle the stronger the correlation; and by the color, red indicates a negative correlation while blue indicates a positive correlation. For example, high scores in Physics are correlated to high scores in Maths, whereas high scores in Physics are correlated to low scores in French.

```
# Run the PCA on the data and plot the variables map
library(FactoMineR)
p = PCA(data32, scale.unit = T, graph = F)
plot(p, choix = "var")
```

Variables factor map (PCA)

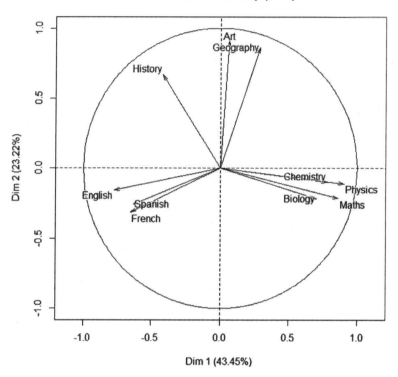

The variables factor map shows the first two PCs or Dimensions, it also shows the percentage of variation explained by these Dimensions, so 43.5% for Dimension 1 and 23.2% for Dimension 2, therefore the first two Dimensions account for 66.7% of the variation.

This plot is also useful for trying to determine what the Dimensions represent. For example, looking at Dimension 1 we can see Chemistry, Biology, Physics, and Maths are strongly toward 1 whereas English, Spanish, and French are strongly toward −1. In addition History is weakly toward −1, but Art and Geography are near 0. This Dimension could therefore represent science subjects versus language subjects.

Looking at Dimension 2 shows History, Geography, and Art are strongly toward 1, whereas the other subjects are weakly toward −1. This Dimension could represent more creative subjects versus noncreative subjects.

```
# Print the PCA eigenvalues
p$eig
```

	eigenvalue	% of variance	cumulative % of variance
comp 1	4.34515923	43.4515923	43.45159
comp 2	2.32216780	23.2216780	66.67327
comp 3	1.66011974	16.6011974	83.27447

comp 4	0.67669161	6.7669161	90.04138
comp 5	0.35969445	3.5969445	93.63833
comp 6	0.28062679	2.8062679	96.44460
comp 7	0.18110805	1.8110805	98.25568
comp 8	0.12340070	1.2340070	99.48968
comp 9	0.03092235	0.3092235	99.79891
comp 10	0.02010928	0.2010928	100.00000

Printing the eigenvalues shows the raw values we use to plot the scree plots below. Eigenvalues should be large for the first Dimension, then gradually diminishing. It also helpfully shows the cumulative percentage of variance; here it refers to "comp" for component, which is the equivalent of a PC or Dimension. One way of determining how many Dimensions should be included in the final dataset is when the cumulative variance reaches a certain value, such as the commonly used 80% cutoff.

```
# Print the two scree plots
library(factoextra)
fviz_screeplot(p, choice = "variance")
fviz_screeplot(p, choice = "eigenvalue", geom = "line")
```

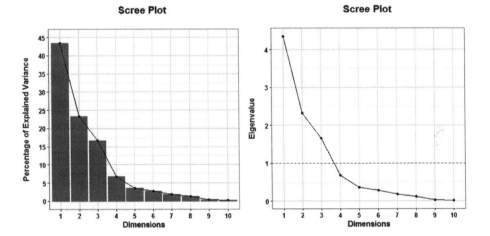

The scree plot on the left shows the percentage of variance explained by each Dimension. Another way of determining how many Dimensions to include is to look at this scree plot and decide where the shape of the curved line changes and tails off, sometimes referred to as a "break."

The scree plot on the right shows the eigenvalues for each Dimension. The third way of determining how many Dimensions to include is to draw a line at 1, and any Dimensions with an eigenvalue larger than 1 should be kept. This is due to the fact that eigenvalues larger than 1 indicate that the Dimension accounts for more variation than one of the original variables.

The three techniques should produce fairly similar answers, maybe only differing by 1 Dimension; it's just personal judgment along with these options to decide how many Dimensions to retain. In this example all three methods agree that 3 Dimensions should be kept. Therefore 3 Dimensions contain 83.27% of the variation of the original 10 variables, a big reduction.

```
# Refit PCA with 3 dimensions and print the variable coordinates
p2 = PCA(data32, scale.unit = T, graph = F, ncp = 3)
p2$var$coord
```

	Dim.1	Dim.2	Dim.3
English	-0.77124549	-0.15716831	0.57766591
Maths	0.86379191	-0.21427753	0.28766160
Chemistry	0.78299456	-0.09753364	0.47112565
Biology	0.70314911	-0.21508135	0.47866252
Physics	0.90376338	-0.11123149	0.25636420
French	-0.65110862	-0.31377262	0.32960466
Spanish	-0.62829343	-0.25407596	0.58367037
History	-0.41639534	0.66881011	0.44793897
Geography	0.28989976	0.86438103	0.03728038
Art	0.06159328	0.90881046	0.27481517

Rerunning the PCA and printing the variable coordinates show the "loadings" of each variable or School Subject. This means how strongly the variable is associated with each Dimension and whether it's a positive or negative relationship.

This information ties into the variables factor map from before, so here we can see the values used to construct this plot. English, French, and Spanish have a strongly negative relationship (-0.77, -0.65, and -0.63) whereas Maths, Chemistry, Biology, and Physics have a strongly positive relationship (0.86, 0.78, 0.70, and 0.90). History has a negative relationship (-0.42), and Geography and Art don't really have a relationship within Dimension 1 (0.29 and 0.06).

They do however have a very strong positive relationship within Dimension 2 (0.86 and 0.91) in addition History has a strong positive relationship (0.67). All the other subjects have quite a weak negative relationship (ranging from -0.10 to -0.31).

Finally in Dimension 3 all the relationships are positive, with English and Spanish having a strong relationship (both 0.58), Chemistry, Biology, and History having a relationship (0.47, 0.48, and 0.45), Maths, Physics, French, and Art having a weak relationship (0.29, 0.26, 0.33, and 0.27) with Geography not really having any relationship (0.04).

```
# Print out the dimension descriptions
dimdesc(p2, axes = 1:2)
```

```
$Dim.1
$Dim.1$quanti
```

	correlation	p.value
Physics	0.9037634	1.439630e-09
Maths	0.8637919	5.420262e-08
Chemistry	0.7829946	6.096205e-06

```
Biology           0.7031491  1.269044e-04
History          -0.4163953  4.297043e-02
Spanish          -0.6282934  1.010126e-03
French           -0.6511086  5.692025e-04
English          -0.7712455  1.025939e-05
```

```
$Dim.2
$Dim.2$quanti
               correlation       p.value
Art              0.9088105  8.153942e-10
Geography        0.8643810  5.182722e-08
History          0.6688101  3.528839e-04
```

dimdesc(p2, axes = 2:3)

```
$Dim.2
$Dim.2$quanti
               correlation       p.value
Art              0.9088105  8.153942e-10
Geography        0.8643810  5.182722e-08
History          0.6688101  3.528839e-04
```

```
$Dim.3
$Dim.3$quanti
               correlation       p.value
Spanish          0.5836704  0.002751891
English          0.5776659  0.003115670
Biology          0.4786625  0.017970219
Chemistry        0.4711257  0.020135919
History          0.4479390  0.028156977
```

The dimension descriptions just highlight which School Subjects are associated with which Dimensions. So at the top it does show that History should be included in the negative relationship of Dimension 1, as it is significant to that Dimension.

For Dimension 3 it does not include the weak relationships of Maths, Physics, French, and Art. In addition you can see the p-values for the remaining subjects, although still significant are larger than previous Dimensions. This Dimension is quite difficult to interpret, as there's no clear linking of the subjects said to be significant. This can happen quite often on the lower Dimensions. However, the first two Dimensions account for a lot of the variation and do lend themselves to a reasonable interpretation, so there is less worry about this Dimension, Dimension 3.

```
# Plot the first three principal components
loadings = as.data.frame(p2$var$coord)
loadings$var = colnames(data32)
library(ggplot2)
ggplot(loadings, aes(x = var, y = Dim.1)) +
        geom_bar(stat = "identity")
ggplot(loadings, aes(x = var, y = Dim.2)) +
        geom_bar(stat = "identity")
ggplot(loadings, aes(x = var, y = Dim.3)) +
        geom_bar(stat = "identity")
```

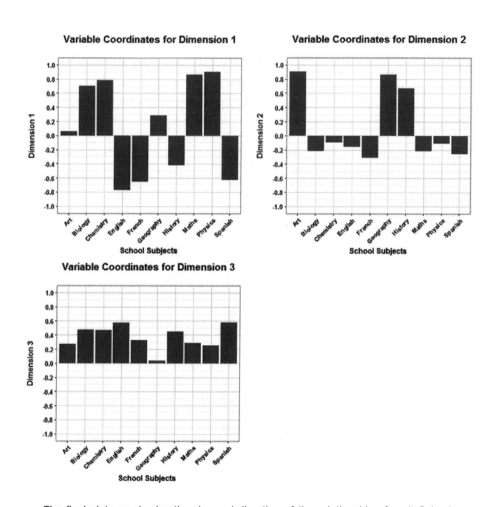

The final plots emphasize the size and direction of the relationship of each School Subject within each of the chosen Dimensions.

As you can see, interpretation of PCA output is very subjective, the data shows which variables have loaded on which components, but defining what those components represent can be quite tricky.

However PCA does condense the data down into a more manageable size, in this example three components are enough to describe the variation from the 10 original variables. This is much more useful with larger data sets and variables that aren't so obviously linked. It was no surprise that the PCA grouped the scientific school subjects and so forth.

Q Methodology

Q methodology is used to group people's opinions into factors, as opposed to grouping variables as with PCA.

This analysis can be used to prioritize areas of funding, capabilities, and so forth or alternatively to gather opinions on a topic (such as new training) to determine whether there are different thoughts from different "populations."

A set of statements are created, called a Q-set, and participants can agree with, disagree with, or be neutral toward these statements. These statements must be representative of the overarching question, aim of the study, but should cover smaller issues linked to that question. It's a good idea to hold panels to decide which statements should be used. The reason for this is people will interpret things differently so the clearest sentence structure should be formed, as well as others potentially pointing out gaps in the statements, so new statements can be created.

Once the final set of statements has been completed a Q-grid needs to be created. These will be triangles for the participants to fit the statements into with the furthest left being negative, the statements they disagree with; to the furthest right, which will be positive, the statements they agree with. Figure 8-1 shows some example Q-grids for 16 statements, 36 statements, and 22 statements.

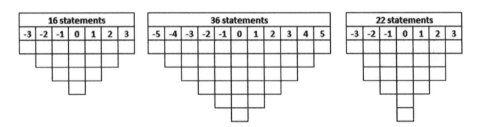

Figure 8-1. Examples of Q-grid

These Q-grids can have a longer tail in the middle or can be wide and thin depending on what you are looking for. The grid should remain symmetrical and should be roughly triangular in shape.

The Q-sort is the title given to the process of the participants ordering their statements; this can be done by hand or on a computer. Initially they should sort their statements into three piles, statements they agree with, disagree with, and are neutral about. Personally I think it's better if it's defined that these three piles should be roughly equal, this makes the next stage of the Q-sort easier.

The participants will then take a pile, say the statements they agree with, and put those into the right side of the triangle. The furthest right cell(s) will be the statement they agree with the most, and then they should work their way left until that have no more statements they agreed with. They should then take the opposite pile, the statements they disagree with, and do the same thing but this time starting from the furthest left cell(s), which would be the statement they disagree with most. Finally they should go through the neutral pile filling in the blank spaces. At any point they can re-jig their Q-sort until they are completely happy with their final response.

The data is collected and the Q method is run on the data. Participants that gave similar rankings in the Q-sort will load significantly on the same factor; these factors can be thought of as perspectives or viewpoints. The output will split the data into factors then show Q-sorts that are most representative of each perspective, however some people may not load onto a factor if the way they have done their Q-sorts fits into multiple factors, though this is generally only a small number of people.

As with PCA the number of factors to retain is a subjective decision, however the same guidelines apply regarding eigenvalues larger than 1, cumulative variance around 80%, or a "break" in the tail. The problem here with eigenvalues is that they will be inflated for big data, however Q methodology wouldn't be used with a large sample size as it was designed for small to medium sample sizes. Additional pieces of information to use for determining the number of factors to retain are that a decent number of people have loaded onto the factor, and that there is a good number of distinguishing statements for all the factors—more on the latter in Example 8.3.

Using the consensus and distinguishing statements you can start to form a picture around the perspectives of different groups. Quite often demographic information is collected from the participants as this can sometimes help explain why certain groups of people have done their Q-sorts in a similar manner, though this isn't guaranteed.

The R package to run the Q methodology is called **qmethod**, it does contain an example data set called **lipset** regarding the values patterns of democracy that has 9 Q-sorts and 33 statements, so you can see that this analysis can be run on a small sample size. The dummy data we use in Example 8.3 concerns effective working, our overarching question is "which factors are the most important to you being happy and effective at work" and the reason we want this question answered is to determine which areas the business should focus on, although this may be different for different groups.

Our final set of 16 statements is as follows:

1. Good pay grades in line with equivalent roles in other companies.

2. Good pension scheme.

3. Ability to work flexible hours.

4. High amount of annual leave days.

5. Interesting work.

6. Up-to-date IT and software.

7. Independent working.

8. Hot desking.

9. Availability of relevant training opportunities.

10. Ability to publish papers on the work undertaken.

11. Management being open to advice and ideas.

12. Involvement in decisions made by senior management.

13. Feeling safe to challenge disagreeable behaviors.

14. Working at an organization that embraces diversity.

15. Affordable catering facilities being provided.

16. Establishing friendships with colleagues.

We asked 10 participants to sort those statements into the first grid shown in Figure 8-1 in order of importance answering the overarching question. So instead of agreeing and disagreeing, they had piles of *very important, not important,* and *neutral.*

EXAMPLE 8.3

The 10 Participants sorted 16 Statements into Q-sorts answering the question "which factors are the most important to you being happy and effective at work." We also collected some demographic data; whether they were in a technical or nontechnical role, whether they had dependents, and what stage of their career they were at: graduate, nonsenior role, or senior role.

```
# Input the data
S1 = c(0,1,2,3,2,1,-1,0,-2,0,0,1,-1,-1,-3,-2)
S2 = c(1,0,2,2,3,1,-2,0,0,-2,-1,-1,-1,-3,0,1)
S3 = c(1,-1,3,1,2,-2,0,0,-1,-3,0,1,0,-1,-2,2)
S4 = c(1,-2,1,0,3,2,-1,-2,0,0,0,2,-1,-1,-3,1)
```

```
S5 = c(0,0,0,-1,1,-2,1,3,0,-3,1,2,-1,2,-2,-1)
S6 = c(3,0,2,1,2,-2,1,-1,0,0,-1,-2,1,-1,-3,0)
S7 = c(2,-1,3,1,0,-2,0,-1,1,-3,0,1,0,2,-2,-1)
S8 = c(2,0,1,2,3,0,-1,-2,-1,-2,0,1,-1,-3,0,1)
S9 = c(1,-2,1,0,2,0,-1,-1,2,1,-1,-2,-3,0,0,3)
S10 = c(2,-1,1,1,3,-1,0,-2,0,0,0,-3,-1,-2,1,2)
data33 = data.frame(S1,S2,S3,S4,S5,S6,S7,S8,S9,S10)
```

```
# Plot the initial data - need to reshape the data to "long" format
data34 = reshape(data33, varying = list(1:10), idvar = "Statement",
        timevar = "Participant", direction = "long",
        v.names = "Response")
data34$Participant = factor(data34$Participant)
library(ggplot2)
ggplot(data34, aes(y = Response, x = Participant,
        fill = Participant)) + geom_bar(stat = "identity") +
        facet_wrap(~Statement) + theme_bw()
```

Response Scores by Statement

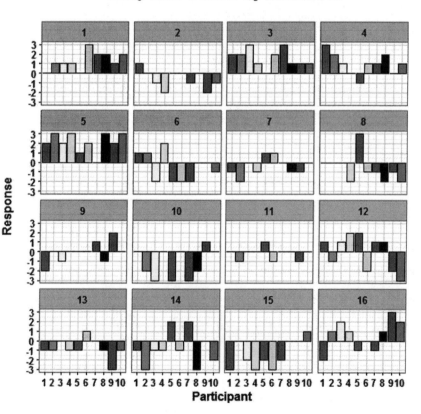

The initial plot looks at the distribution of Responses for each Statement, you can start for form a picture of which Statements people agreed and disagreed with, in general and in comparison to each other. For example, people seemed to agree with each other on Statement 5, interesting work was important. Whereas they disagreed more with each other on Statement 12, involvement in decisions made by senior management was important to some people and not important to others, in comparison to the other Statements.

```
# Run the Q-methodology on the original data and print the summary
library(qmethod)
qm = qmethod(data33, nfactors = 3, rotation = "varimax")
summary(qm)
```

```
Q-method analysis.
Original data:              16 statements, 10 Q-sorts
Forced distribution:        TRUE
Number of factors:          3
Rotation:                   varimax
Flagging:                   automatic
Correlation coefficient:    pearson
```

Factor scores

	fsc_f1	fsc_f2	fsc_f3
1	1	2	2
2	0	-1	-1
3	2	1	3
4	2	0	1
5	3	3	1
6	1	-1	-2
7	-2	0	0
8	0	-2	0
9	-1	1	0
10	-1	0	-3
11	0	0	0
12	1	-3	1
13	-1	-2	-1
14	-3	-1	2
15	-2	0	-2
16	0	2	-1

	f1	f2	f3
Average reliability coefficient	0.80	0.80	0.80
Number of loading Q-sorts	4.00	2.00	3.00
Eigenvalues	2.89	2.56	2.21
Percentage of explained variance	28.87	25.62	22.07
Composite reliability	0.94	0.89	0.92
Standard error of factor scores	0.24	0.33	0.28

Due to the number of Participants 3 factors were chosen for the Q method. The output shows the most representative Q-sorts for each of the 3 factors; so this highlights which statements were positively or negatively associated with the factors and also the strength of that association. The plot below, created from the first output, shows the top 3 most important statements in shades of green (5, 3, and 1) and the 3 least important statements in shades of red (14, 12, and 10). You can see that the greens stay fairly high across all three factors, whereas the reds are split across the Q-sorts for the different factors. It is likely that the "perspectives" were split into three more so from the "negative" statements, the statements of things that weren't important to people, as there seems to be a fair amount of agreement on the most important things.

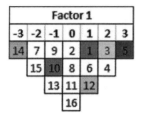
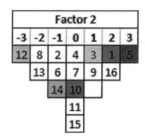
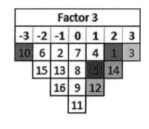

The next set of outputs show that all three factors have eigenvalues larger than 1 and show the percentage of variation explained by each factor, totaling 76.56%. It also shows the number of Participants that loaded onto each factor, 4, 2, and 3, respectively. This means that one Participant didn't load onto any factor.

```
# Print the Q-method loadings and flagged Q-sorts
qm$loa
```

	f1	f2	f3
S1	0.86152724	-0.23766219	0.169804927
S2	0.75022766	0.48389174	0.035115411
S3	0.49682774	0.29219127	0.713349044
S4	0.69460242	0.23480173	0.143712479
S5	-0.06938019	-0.40538253	0.760998161
S6	0.31608729	0.48960577	0.513287267
S7	0.13835638	0.05971983	0.888139998
S8	0.80138354	0.41395274	0.124590881
S9	0.06212871	0.85144371	-0.016674981
S10	0.28695064	0.90922804	-0.009836156

```
qm$flagged
```

	flag_f1	flag_f2	flag_f3
S1	TRUE	FALSE	FALSE
S2	TRUE	FALSE	FALSE
S3	FALSE	FALSE	TRUE
S4	TRUE	FALSE	FALSE
S5	FALSE	FALSE	TRUE
S6	FALSE	FALSE	FALSE
S7	FALSE	FALSE	TRUE
S8	TRUE	FALSE	FALSE
S9	FALSE	TRUE	FALSE
S10	FALSE	TRUE	FALSE

The loadings output shows which factor each Participant loaded on and the strength of that loading. Looking at the flagged output does the work for you and shows you exactly which Participant loaded on which factor. So Participants 1, 2, 4, and 8 loaded on Factor 1, Participants 9 and 10 loaded on Factor 2, and Participants 3, 5, and 7 loaded on Factor 3; Participant 6 didn't load on any factor.

The demographics data was as follows, sorted by factors:

Factor	Participant	Role	Dependent(s)	Career Stage
1	1	Technical	Y	Senior
	2	Technical	Y	Senior
	4	Technical	N	Senior
	8	Technical	N	Non-senior
2	9	Technical	N	Graduate
	10	Technical	N	Graduate
3	3	Non-Technical	Y	Senior
	5	Non-Technical	N	Senior
	7	Non-Technical	Y	Non-senior
-	6	Technical	Y	Non-senior

I've highlighted some patterns in the demographics; Factor 2 separates the graduates' perspective from the senior and nonsenior Participants, Factor 1 separates the remaining technical Participants, and Factor 3 the remaining nontechnical Participants. Apart from the graduates the career stage doesn't seem to have any impact on the perspectives, nor does having any dependents. This could be useful information to take forward when thinking about areas to focus on within groups.

```
# Print the distinguishing and consensus statements
# Values rounded for space saving
qm$qdc
```

```
              dist.and.cons    f1_f2 sig    f1_f3 sig    f2_f3 sig
1                   Consensus   -0.4279       -0.2988       0.1290
2                                0.9619 *      0.5996      -0.3624
3                                0.5169       -0.4519      -0.9688 *
4   Distinguishes f1 only        1.1263 **     1.1721 **    0.0458
5   Distinguishes f3 only        0.2043        1.4880 **    1.2836 **
6       Distinguishes all        1.0807 **     2.0985 **    1.0178 *
7                               -0.6449       -1.0635 **   -0.4186
8   Distinguishes f3 only        0.4600       -0.7327 *    -1.1927 **
9   Distinguishes f1 only       -1.2570 **    -1.0311 **    0.2259
10      Distinguishes all       -0.9250 *      1.4698 **    2.3947 **
11                Consensus      0.0976       -0.3210      -0.4186
12 Distinguishes f2 only         2.2989 **    -0.3272      -2.6261 **
13                               0.4079       -0.5690      -0.9769 *
14 Distinguishes f3 only        -0.5918       -2.4373 **   -1.8455 **
```

```
15 Distinguishes f2 only  -1.6230 **    0.2255      1.8484 **
16 Distinguishes f2 only  -1.6849 **    0.1792      1.8641 **
```

The distinguishing and consensus statements output shows for each statement whether there were significant differences between the factors, an asterisk (*) is at 95% confidence and two asterisks (**) is at 99% confidence.

There were only two statements that "distinguishes all," which means all factors were significantly different to each other in the perspectives, for Statement 6 (up-to-date IT and software) and Statement 10 (ability to publish papers on the work undertaken). "Distinguishes f1 only" means that Factor 1 was significantly different in perspective to Factor 2 and Factor 3, but there was no difference between Factor 2 and Factor 3. Consensus means that all three factors were in agreement with the statement, so there was consensus in perspective for Statement 1 (good pay grades in line with equivalent roles in other companies) and Statement 11 (management being open to advice and ideas). Sometimes the consensus statements can be just as important as the distinguishing statements.

```
# Plot the z-scores
plot(qm, legend = T)
```

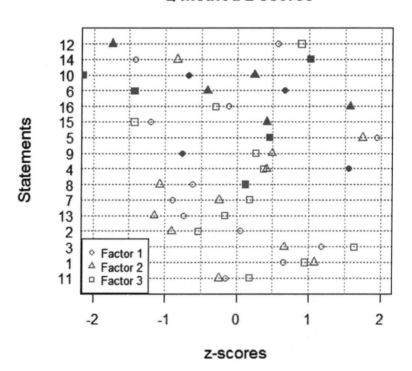

Q-method z-scores

Plotting the z-scores is a really useful way of visualizing the output from the distinguishing and consensus statements. It orders the statements in terms of most differing perspectives to least differing perspectives, so you will always have the consensus statements at the bottom, however you may not always have the "distinguishes all" statements at the top.

In this plot the perspective of the graduates for Statement 12 (involvement in decisions made by senior management) is so different to that of everyone else that it's at the top of the plot, above the "distinguished all" statements. By the point being out to the left, a negative z-score means that this statement was one of the least important to the graduates.

The points are filled when it's involved in a distinguishing statement, which is why the point for Factor 2, the graduates, is filled next to Statement 12 as the distinguishing and consensus output tells us that Statement 12 "distinguishes f2 only."

As mentioned before, the consensus statements can be just as important. Although there was consensus for Statement 11 (management being open to advice and ideas), this was a neutral choice, whereas the consensus for Statement 1 (good pay grades in line with equivalent roles in other companies) was a more positive, and therefore important, choice for the Participants.

By combining all the output we can see that the items of most importance to technical staff are interesting work, flexible hours, and annual leave (Statements 5, 3, and 4), and the items of least importance are organization embracing diversity, independent working, and affordable catering (Statements 14, 7, and 15). For graduates the most important items are interesting work, good pay, and friendships (Statements 5, 1, and 16), and the least important items are involvement in senior managers decisions, up-to-date IT and software, and challenging disagreeable behaviors (Statements 12, 6, and 13). For nontechnical staff the items of most importance are flexible hours, good pay, and embracing diversity (Statements 3, 1, and 14), and the items of least importance are publishing papers, up-to-date IT and software, and affordable catering (Statements 10, 6, and 15).

If you were restricted in what areas could be focused on, then using the representative Q-sorts and the z-scores plot, the things recommended to take forward would be interesting work, flexible hours, good pay, and annual leave, although this is a subjective judgment. It may be that the organization has no control over the hours, pay, and leave, in which case while you may know it will have some effect on the happiness and efficiency of the staff, it isn't an area the business can focus on improving, which is the aim of the study. As such statements like that shouldn't be included in the original Q-set, which is why careful consideration needs to be given to the initial set of statements.

As with PCA, interpretation of Q methodology output is very subjective. Sometimes it won't be clear why different people have grouped together in the way that they have, unlike our example where the demographics showed a link between them.

In an ideal situation there would be consensus among everyone and there would only be one representative Q-sort, however in practice this is unlikely to happen. It's up to you to determine how different the Q-sorts are on each factor to decide whether different action should be taken for each group or whether there is enough consensus that you can consider just one approach.

The example highlights how useful the z-scores plot is for visualizing the distinguishing and consensus statements of each group of people. It's also handy to create Q-sorts from the representative Q-sorts to look for common patterns with the statements.

Summary

This chapter extends the statistical modeling into data that has multiple response variables. It has looked at a very small selection of multivariate methods, however they have been chosen due to being dissimilar enough with regard to the data type and the aim of the question, as well as being arguably the most common methods to use in those cases. It also briefly mentioned some other multivariate techniques that can be looked up if more information is required.

The first technique investigated was multivariate analysis of variance (MANOVA), which is the most similar to univariate statistical modeling methods. It is still concerned with looking for significant differences between explanatory variables, but uses two or more dependent response variables. This section also described the assumptions related to conducting MANOVA testing and what to do if the assumptions are violated.

The example given used shot data, the response variables were the X and Y Coordinates, and the explanatory variables were Weapon and Distance. The example highlighted how to test the assumptions, run the MANOVA, run multiple comparisons tests, and plot and interpret the results.

The second method examined was principal component analysis (PCA), which is used when there are many correlated response variables with or without some independent explanatory variables such as a grouping variable. It described how PCA is used to compress the original variables into a smaller subset of components losing as little information as possible. It described the limitations of PCA, and also highlighted how it's more of an exploratory step before then using the components found in further analysis, as well as being highly subjective.

The example presented looked at finding a relationship between the test scores of participants in 10 school subjects. It showed how to run the PCA on the data, how to determine the best number of components to retain, how to see which variables have loaded on which components, and how to plot and interpret the final results.

The third and final technique described was Q methodology, which is used when there are many subject viewpoints that need to be grouped to find overall perspectives. It described how to collect the data, by creating a set of statements, Q-set; and a preset grid, Q-grid; and asking participants to sort those statements into the grids, Q-sort. It then gave some information about what the Q method output will be able to tell you, as well as the fact that it is useful to collect demographic data from the participants to potentially be used with the Q method output.

The example looked at the Q-sorts of 10 participants answering a question about happiness and efficiency in the workplace using 16 statements. It showed how to run the Q method on the data, how to determine the final number of factors to retain, how to see who has loaded on to which factors, how to understand the distinguishing and consensus remarks, and how to plot and interpret the results.

Chapter 9 concentrates on how to present useful, clear graphs to the customer to visually highlight any results found in simple hypothesis testing, statistical modeling, or multivariate data analysis. It focuses on which aspects keep a plot simple but effective, as well as highlighting common mistakes that just confuse or detract from the key message.

The examples are run in R showing the useful bits of code for changing structure, color, axes, labels, and general plot aesthetics. The graphs are those used in previous chapters to highlight the simple tweaks to take a basic plot to one that clearly shows all the information that needs to be shown, in a format that can be included in final reports, papers, or presentations.

Graphs

What Does the Data Look Like?

The best way to convey a key message is through visualization, so picking the most clear and concise graph to do that is crucial.

Contrary to popular belief, using Excel to plot data isn't necessarily the easiest software to use or the most appropriate. Once you understand the **ggplot()** code in R, it is very easy to create simple but informative graphs for all different data types.

Chapter 3 described which plots could be drawn for which data types, so I will not repeat that here. However more detail will be given as to how to improve the look of plots in R along with some general advice about what makes a better graph and what to avoid.

The first section looks at some general things to avoid doing when plotting graphs; the next discusses the commonalities between all the **ggplot()** codes, such as creating titles, axis limits, and so forth; and the final section illustrates some R code that produces the different types of plots, including code in R that doesn't use the **ggplot2** package.

Common Plotting Mistakes

There are many common graphs but not all are suitable for all data types as discussed in Chapter 3. In addition, more thought and detail should be given to the plots that will be presented to the customer as they can convey a lot of information. However, the more impressive the plot does not always mean it is better; it still should be kept as simple as possible while still conveying the same message.

©Victoria Cox 2017
V. Cox, *Translating Statistics to Make Decisions*, DOI 10.1007/978-1-4842-2256-0_9

3D and Pie Charts

One of the most commonly misused feature with plots is 3D. In most cases making the graph 3D will not add any extra information and in fact will probably make the plot harder to interpret.

Linked to this is a pie chart, especially a 3D pie chart, as they are very common plots yet are one of the most difficult to comprehend. There is nothing that can be plotted in a pie chart that couldn't be plotted, more clearly, in a bar chart. In addition to this, if a 3D pie chart had been created, then depending on the angle of the pie chart, the segment sizes seem different visually. Figure 9-1 shows two rotations of a 3D pie chart and the same information in a bar chart.

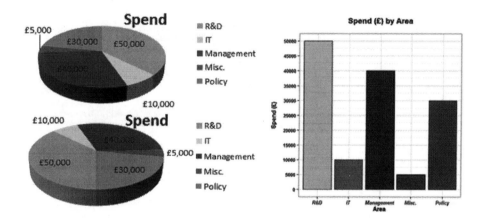

Figure 9-1. 3D pie charts

In the first 3D pie chart the Management section appears larger than the R&D section even though it is a lower number. If the 3D pie chart is rotated 180° the Management section looks smaller than the R&D section, except now the Policy section looks a similar size to the Management section.

The bar chart shows the same information but is much clearer, there's no confusion about which spend-by-area is greater or smaller than the others; if more detail was required about the exact amounts these could either be added on to the bars or shown simply in a table.

There will be cases where a 3D plot will be necessary, for example, when there are three continuous variables, see Figure 9-2.

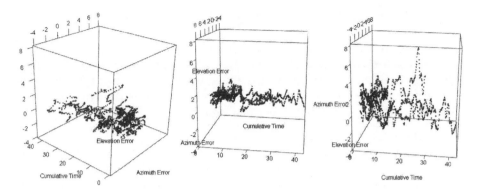

Figure 9-2. 3D plot with rotation

These plots show the errors in the *x* and *y* coordinates, azimuth and elevation, over time. The plot itself can be rotated within R to get the best view, whereas I have chosen three rotations of the plot, to show in the flat figure, to highlight the different views that can be seen. This plot was drawn using **plot3d()** within the **rgl** package. It also could have been plotted using **scatterplot3d()** within the **scatterplot3d** package, however this latter plot cannot be rotated manually, see Figure 9-3.

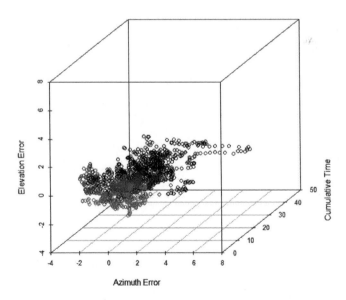

Figure 9-3. 3D plot with no rotation

In this plot the red is lighter at times closer to 0 and darker at times closer to 50.

Plotting Averages

A common mistake is to plot averages using a bar chart; this should be avoided as bar charts should be used for total counts or percentages. A bar implies "this whole amount" whereas an average is exactly that, an average. There are data values above the average point as well as below the average point, but a bar chart will only color the values below, which is misleading.

In situations like this, a dot plot or a line graph should be created, preferably with error bars. For example, if you had multiple subjects that showed a similar trend, such as with the heart rate data in Chapter 7, Example 7-10, then a line graph with error bars would be a good representation of the data, see Figure 9-4.

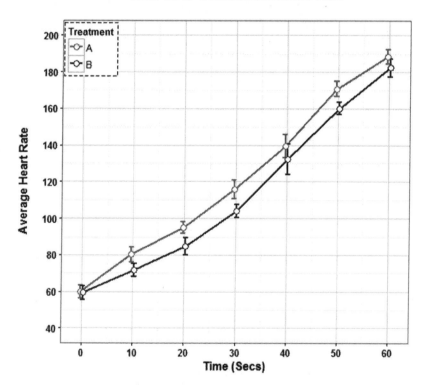

Figure 9-4. Line graph with error bars

It is not always preferable to calculate the average of the data, sometimes it would be worth plotting the raw data to show the variation between groups. For example, if you had different trends between categorical groups you wouldn't want to just plot an average with error bars. Figure 9-5 shows the average data, the average data with error bars, and the raw data for Temperature by Day.

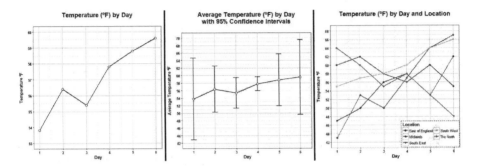

Figure 9-5. Comparison of line graphs

It's worth noting that the average data without error bars is very misleading as it suggests an upwards trend. It is also on a different y-axis scale to the other two plots and doesn't state that the data shown are averages.

Adding in error bars, here 95% confidence intervals, reduces the appearance of a positive trend somewhat and shows there is large variation within the data.

Finally, the last plot gives all the information, as it shows that there are different trends for different locations, so an average would not have been appropriate even with error bars.

Multiple Plots

If you have multiple explanatory variables with multiple levels, a lot of plots are limited by the number of variables that can be shown in one go—therefore the need for multiple plots. However this takes up a lot of space, separates the information up when it may not be independent, and means you may miss visual patterns and trends.

The way around this is to use trellis plots that can display a lot of information from multiple variables without taking up too much space and without overcrowding a plot. Although this obviously has limitations, you should use common sense when deciding on this.

A visual example has not been shown here as there is an adequate plot in Example 9.1 in the next section, the second of the two plots. This plot shows four explanatory variables, Distance, Target Size, Operator, and Ammo Type with 6, 3, 3, and 2 levels, respectively against a response variable of Accuracy. This information is still very clear to see, however the decision was made not to include error bars, though there were multiple repeats per point, as this would have overcrowded the plot unnecessarily.

Another advantage is to swap the order of the explanatory variables; it can change the emphasis within the plot. If the analysis showed differences or trends between two of the four variables, the plot can be amended to display that more clearly. The trellis plot in Example 9.1 easily highlights any key messages from the statistical analysis undertaken.

Plotting Ordinal Data

Ordinal data is frequently treated incorrectly in both analysis and graphs; the data has a natural ordering and should be treated as such. The data should never be used as continuous, so no averages.

Bar charts could be used as essentially you are counting how many of each response there are, but there is a much clearer plot that can be used: a Likert plot, see Figure 9-6.

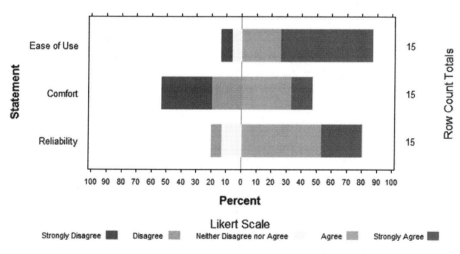

Figure 9-6. Likert plot for ordinal data

These plots clearly show the order of the responses as well as coloring them appropriately. In addition the scales can be shown in percentages, as above, while also displaying the sample size for each group, which is on the right in this example.

Open Text Responses

When surveys have open text boxes data can still be highlighted visually using word maps. In R you can use the ***tm, wordcloud***, and ***SnowballC*** packages to create attractive word maps.

There are many options along the way to remove punctuation, spaces, and more important stop words, such as *the, it, and*, and so forth. You also can add in color and size dependent on the frequency of words, see Figure 9-7.

Figure 9-7. Word map

One issue that can arise by using the different commands to whittle the data down to "useful" terms is that the end of the words can be chopped off. In the example above a few words have lost the e, such as update, awesome, and please.

Unnecessary Plots

In the majority of cases a plot will display information much clearer than general text or a table, however there are some cases where this is not the case. Plots shouldn't be drawn for the sake of it.

In some cases it may be preferable to have a table with figures alongside a plot, but it also may be preferable to have a table instead of a plot when there isn't much data. For example, consider the counts of respondents that smoke or not split by gender—this information boils down to four numbers. Figure 9-8 shows a plot and a table of the results.

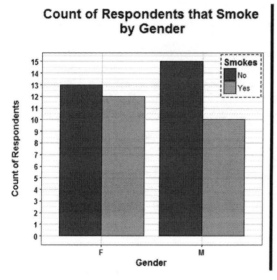

Respondents that Smoke by Gender		Smokes		Total
		Yes	No	
Gender	F	12	13	25
	M	10	15	25
Total		22	28	50

Figure 9-8. Plot versus tables

In this case the plot is fairly redundant and takes up more space than the table. If there were more explanatory variables, or more groups within a variable such as more than Yes or No for the smoking question, then a plot may be more desirable. However in this case the table will suffice and be clear enough without the plot.

Display

The term *display* here covers all manner of details, but they are all very important to the effectiveness of the plot. Once the correct plot has been chosen, the main things to think about are labels, colors, shapes, and fonts.

Make the labels as clear and concise as possible; units can be a good piece of information to include next to an axis title, especially if the data had to be logged. You should exclude the type of plot from the title, it will be clear which plot you have chosen or you can mention it in the text. Make sure any abbreviations of variable levels or variables themselves are extended for the plot, unless this makes the text too long in which case a key will be required.

Adaptation of color may be required dependent on the media to be used or the audience. R has many shades of gray if color can't be seen in the chosen media. In addition some people may be color blind, which requires avoiding certain combinations of colors, there is a list of all the colors you can choose in R.[1]

Linked to color is shape, whether this is different shapes for points or different line types. There are many options that can be combined with different colors to reduce the risk of a viewer not being able to differentiate groups for any reason.

Finally the font, this applies to the font itself as well as the text size, whether the text is bold, italics, and so forth. Both the labels and the numerical axes should be clear, concise, and a decent size in relation to the plot itself. A good rule-of-thumb is to have the title in the largest size, the axes labels and any legend or facet labels the next size, then finally the axes check marks—whether numerical or not, the smallest size while still being legible.

Graph Aesthetics in R

This section moves on to the actual code for plotting graphs in R using the **ggplot2** package. Although it may seem contradictory to start with the details of creating a plot in R, this is because the aesthetics are common across all the different **ggplot()** graphs and a good plot can't be created without sorting these details out first.

Example 9.1 shows a basic plot without the extra aesthetics, then another including the extra aesthetics. After the example, each section of the code is detailed but not all available options have been included for reasons of space.

If you are not interested in the R code then skip to the next section after viewing the example.

---

[1]`http://www.stat.columbia.edu/~tzheng/files/Rcolor.pdf` is invaluable for picking colors in R.

EXAMPLE 9.1

This example uses the data from Figure 3-10 in Chapter 3 that shows a trellis graph.

```
# Input the data
Accuracy = c(37.8,35.2,39.1,38.4,39.6,40.9,33.4,32.1,34.8,35.7,34.6,
          35.7,33.1,32.5,32.4,33.6,33.9,34.6,31.7,31.9,30.8,32.1,32,
          31.4,30.9,31,30.7,31.1,31.8,32.6,28,28.6,27.7,28.3,29,28.4,
          45.3,42.8,47.5,49.7,46.3,49.8,39.8,39.2,39.7,40.2,40.1,41,
          37.1,35.3,39.2,38.4,39.6,40,36.2,35.6,35.7,36,36.3,37.1,33.6,
          33.8,33.9,33.7,34.9,35.8,31.3,31.6,31.8,31,32,31.4,49.8,47.1,
          50.8,53.4,51.7,54.2,45.6,44.3,47.9,43.7,49.8,53.9,42.1,42.9,
          41.5,41.9,42.9,43.6,39.1,38.5,38.4,38.7,39.5,40.5,38.9,38.2,
          38.7,39,37.8,37.1,35.3,35.4,35.2,35.2,35.7,36)
Target = rep(c("Small","Medium","Large"), each = 36)
Ammo = rep(c("A","B"), 54)
Operator = rep(c("Op1","Op2","Op3"), each = 2, 18)
Distance = rep(c(1:6), each = 6, 3)
data35 = data.frame(Target, Ammo, Operator, Distance, Accuracy)

# Plot a basic plot
library(ggplot2)
ggplot(data35, aes(x = Distance, y = Accuracy, colour = Target)) +
     facet_wrap(~ Ammo + Operator) + geom_point()
```

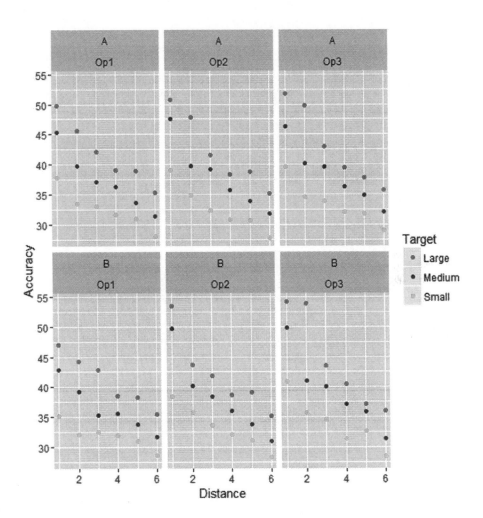

The basic plot uses the variables as axis labels, cannot create a title, calculates the x-axis and y-axis limits and check marks, and adds labels; and generally doesn't look very pretty.

```
# Plot a better presented plot
library(ggplot2)
ggplot(data35, aes(x = Distance, y = Accuracy, colour = Target)) +
        theme_bw() + facet_wrap(~ Ammo + Operator,
                labeller = label_wrap_gen(multi_line = FALSE)) +
        geom_point() + xlab("Distance (m)") + ylab("Accuracy\n") +
        ggtitle("Accuracy by Distance, Ammo Type,
                Operator and Target Size") +
        scale_colour_manual("Target", values = c("firebrick3",
                "forestgreen","dodgerblue3")) +
        scale_x_continuous(limits = c(1,6), breaks = seq(1,6,
```

```
          by = 1)) +
scale_y_continuous(limits = c(25,55), breaks = seq(25,55,
          by = 5)) +
theme(plot.title = element_text(size = 16, face = "bold"),
  strip.text = element_text(size = 12, face = "bold"),
  legend.background = element_rect(linetype = "dashed",
          colour = "black", fill = "grey85"),
  legend.key = element_rect(colour = "grey30", size = 0.5),
  legend.position = "top",
  legend.title = element_text(size = 12, face = "bold"),
  legend.text = element_text(size = 10),
  axis.text = element_text(size = 9, face = "bold"),
  axis.title = element_text(size = 12, face = "bold"),
  panel.grid.major = element_line(colour = "grey90"))
```

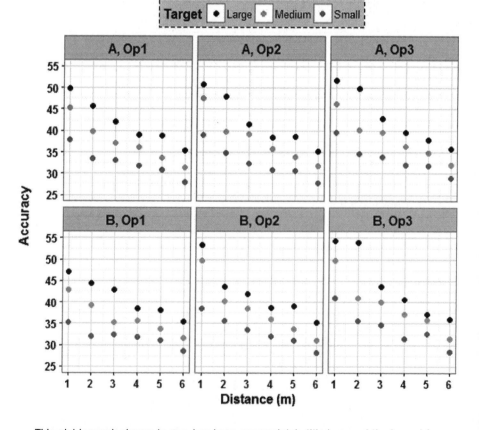

This plot is much clearer to see, has been appropriately titled, moved the legend for better spacing, improved the colors, and more.

Breaking the initial ggplot code into its components:

- **data** – this is the data set you are working with.

- **x = Distance** – this is the explanatory variable you want on the x-axis/horizontal axis, here we used Distance.

- **y = Accuracy** – this is the response variable to be put on the y-axis/vertical axis, here we used Accuracy.

- **colour = Target** – this is an optional explanatory variable you want to show using different colors, here we picked Target.

- **facet_wrap(~ Ammo + Operator)** – this is the "panels" for each level of the additional explanatory variables you supply, here we used Ammo and Operator.

- **geom_point()** – this let's R know that we want to create a scatterplot.

Now adding on the extra aesthetics:

theme_bw()

This just gives a nicer structure to the overall plot, makes sure the background is white and that the axes and panels have a black line.

facet_wrap(~ Ammo + Operator, labeller = label_wrap_gen(multi_line = FALSE)

This lets you change the position of the "panel" labels to be next to each other and separated by a comma, instead of placed on top of each other as in the original plot, which uses up unnecessary space.

xlab(" "), ylab("\n"), and **ggtitle(" ")**

These let you amend the x-axis and y-axis labels and add a title to the plot. By adding **\n** at the end of the text within a command, a carriage return is left between the bottom of the label and the plot, therefore it is only used with **ylab()** and **ggtitle()**. Instead of adding **\n** you also can put a carriage return in yourself instead, as I did for the plot title in Example 9.1.

scale_colour_manual("Target", values = c("firebrick3", "forestgreen", "dodgerblue3"))

Here you can define which colors to use in the plot, remember you need to list the same number of colors as levels in the chosen explanatory variable. The color would change to fill for bar charts and box plots, and can be replaced with shape to list the different shapes if that has been specified earlier on.

scale_x_continuous(limits = c(1,6), breaks = seq(1, 6, by = 1))

This command lets you set the limits of the x-axis as well as defining where the check labels should be. So here the x-axis will be labeled 1, 2, 3, 4, 5, and 6; it may be that you want more space at the edges and could have the limits as ***c(0.5, 6.5)*** but leave the breaks as they are. The same command is used for the y-axis using ***scale_y_continuous()***.

The next set of commands are all held within ***theme()***; they control the font sizes, appearances, and all details contained within the legend.

Using the commands including ***element_text()***; ***plot.title = element_text()***, ***strip.text = element_text()***, ***legend.title = element_text()***, ***axis.text = element_text()***, and ***axis.title = element_text()*** lets you control all the text in the plot. Within each command you can change the following:

- font size: ***size = 12***, gives a font size of 12.

- Emphasis: ***face = "plain"***, ***face = "italic"***, ***face = "bold"***, and ***face = "bold.italic"***, gives you the four possible options.

- color: ***colour = "black"***, or ***color = "black"***, both would make the text black.

- rotation: ***angle = 45***, gives a 45° rotation to the text but can be any degree between 0 and 360.

- Justification: ***h = 1*** or ***v = 1***, mainly used when a rotation of text has been used, h is primarily used with an x-axis rotation and v is primarily used with a y-axis rotation.

To separate the commands for each axis independently, instead of ***axis.text*** you would specify ***axis.text.x*** or ***axis.text.y***.

The next few lines let us control elements linked to the legend.

legend.position = "top"

This lets you specify the position of the legend. It can either be one of four main positions outside the plot; ***"top," "left," "bottom,"*** or ***"right,"*** or it can be put inside the plot region using ***c(0.2, 0.7)***. You may need to play around with the numbers so that the data isn't hidden beneath the legend.

legend.background = element_rect()

legend.key = element_rect()

The first of these is concerned with the background of the legend as a whole, whereas the second relates to the individual sections of the legend; the labels for the levels of the variable, see Example 9-1. In both cases, and in fact any *element_rect()* commands, the following can be changed:

- border line size: *size = 1*, gives a border line size of 1.

- background fill color: *fill = "grey85"*, gives you a light gray background.

- border line color: *colour = "black"*, or *color = "black"*, both would make the border line color black.

- border line type: *linetype = "solid"*, gives a solid border line, the other main options are *"dashed," "dotted," "dotdash," "longdash," "twodash"*; or the numbers 1 to 6 can be used, respectively.

The last line lets you add grid lines to the plot to make identifying the position of the data easier.

panel.grid.major = element_line(colour = "grey90")

The following can be changed with any *element_line()* commands:

- grid line size: *size = 1*, gives a grid line size of 1.

- grid line color: *colour = "grey90"*, or *color = "grey90"*, both would make the grid line a light shade of gray.

- grid line type: *linetype = "solid"*, gives a solid border line, the other main options are as previous.

As a side note, if you wish to resize the plot you can use the command *windows()*; for example *windows(6,6)* will give a square plot, whereas *windows(7,4)* will give a long thin plot.

The next section looks at the different R commands to plot the graphs using the *ggplot2* package and others where more appropriate than using *ggplot2*. Commands listed earlier for the aesthetics will be used without extra explanation due to the details being given, however they will be common across all *ggplot2* package plots.

Graphs in R

We now look at some of the graphs that can be plotted in R to show a clear message for the customer, this should be done along with relevant text that sums up the output.

Bar Chart

Using the data from Chapter 7, Example 7.6, we can plot a bar chart showing the number of detections that highlights a detection is a good thing using the green color. We also can show what the counts would be in percentages by each terrain within each device in Example 9.2.

The **geom_bar()** is what allows us to create the bar chart, and we have chosen *"identity"* as we want to use the raw values rather than sum up binary counts, *"dodge"* as we want the bars next to each other and not stacked, and *"black"* as we want black lines around the bars.

The **geom_text()** section lets us add the percentage labels to the plot defining exactly how they should be calculated and where they should be placed.

EXAMPLE 9.2

Creating a bar chart in R using data from Example 7.6, data24, as this was created from the table of frequencies.

```
ggplot(data24, aes(x = Terrain, y = Freq, fill = Detection)) +
      geom_bar(stat = "identity", position = "dodge",
              colour = "black") +
      facet_wrap( ~ Device) + theme_bw() + xlab("Terrain") +
      guides(fill = guide_legend(ncol = 2)) + ylab("Detection") +
      scale_fill_manual("Detection", values = c("darkred",
          "forestgreen")) +
      scale_y_continuous(limits = c(0,15), breaks = seq(0,15,
              by = 1)) + ggtitle("Frequency of Detections by
              Terrain and Device\n")+
      geom_text(aes(y = Freq, label = paste(round(Freq/15*100,1),
              "%")), position = position_dodge(width=0.99),
              size= 2.5, vjust = -0.5) +
      theme(plot.title = element_text(size = 16, face = "bold"),
      legend.background = element_rect(linetype = "dashed",
              colour = "black"),
      legend.position = c(0.11,0.91),
      strip.text = element_text(size = 12, face = "bold"),
      axis.text = element_text(size = 10, face = "bold"),
      axis.title = element_text(size = 12, face = "bold"),
      panel.grid.major = element_line(colour = "grey90"))
```

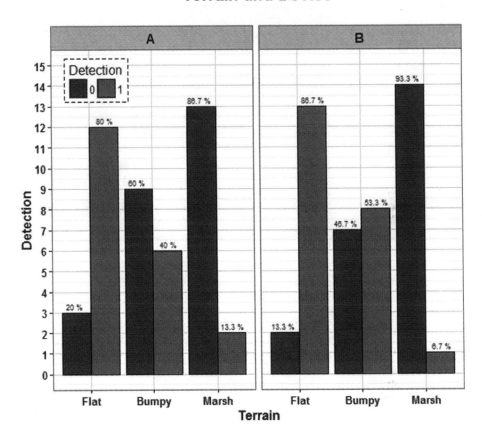

Frequency of Detections by Terrain and Device

These colors can always be amended dependent on the audience and media to be used.

If you are only showing percentages on a bar chart remember to include the sample size of the groups somewhere, as for example 100% can look very impressive until it's highlighted that the sample size was 2.

Tile Plot

A tile plot can be used to display small amounts of discrete data, using the data from Chapter 6, Example 6.3, we can show the split between the paired data of task completion with different suits in Example 9.3. The example also shows how to create a tile plot for larger than 2 × 2 tables using the data from Example 9.2.

The *geom_tile()* is what lets us create the tiles themselves, and we have chosen *"Freq"* as we want it to fill the colors using the frequencies, and *"black"* as we want the tiles to have a border.

The *scale_fill_gradient()* section here lets us define the limits and colors we want to use. If the plot was run without this it would assume the largest frequency was the upper limit and therefore color it with the "top" color, which could be misleading.

The *geom_text()* as before allows us to add the percentage labels to the plot.

EXAMPLE 9.3

Creating a tile plot in R using data from Example 6.3, data11, and from Example 9.2, data24.

```
ggplot(data11, aes(Var2, Var1)) + geom_tile(aes(fill = Freq),
       colour = "black") + theme_bw() +
       scale_fill_gradient(limits = c(0,32),  breaks = seq(0,32,
              by = 4), low = "white", high = "deepskyblue3") +
       xlab("Suit 2") + ylab("Suit 1") +
       ggtitle("Frequency of Task Completion in Each Suit") +
       geom_text(aes(label = paste(round(Freq/32*100,0),"%"))) +
       theme(plot.title = element_text(size = 16, face = "bold"),
       legend.background = element_rect(linetype = "dashed",
              colour = "black"),
       legend.title = element_text(size = 12, face = "bold"),
       legend.text = element_text(size = 9),
       axis.text = element_text(size = 10, face = "bold"),
       axis.title = element_text(size = 12, face = "bold"))
```

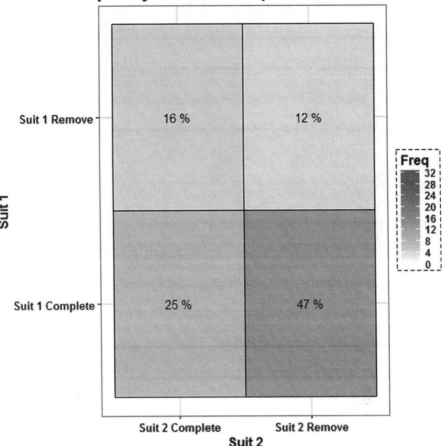

Frequency of Task Completion in Each Suit

This tile plot has been amended from the one in Example 6.3 so that the upper limit of 32 is the darkest color as opposed to the highest frequency.

```
ggplot(data24[data24$Detection == 1,], aes(Terrain, Device)) +
        geom_tile(aes(fill = Freq), colour = "black") +
        scale_fill_gradient(limits = c(0,15), breaks = seq(0,15,
                by = 3), low = "red", high = "forestgreen") +
        theme_bw() + xlab("Terrain") + ylab("Device\n") +
        ggtitle("Frequency of Detections by Terrain and Device") +
        geom_text(aes(label = paste(round(Freq/15*100,0),"%"))) +
        theme(plot.title = element_text(size = 16, face = "bold"),
        legend.background = element_rect(linetype = "dashed",
                colour = "black"),
        legend.title = element_text(size = 12, face = "bold"),
        legend.text = element_text(size = 10, face = "bold"),
        axis.text = element_text(size = 10, face = "bold"),
        axis.title = element_text(size = 12, face = "bold"))
```

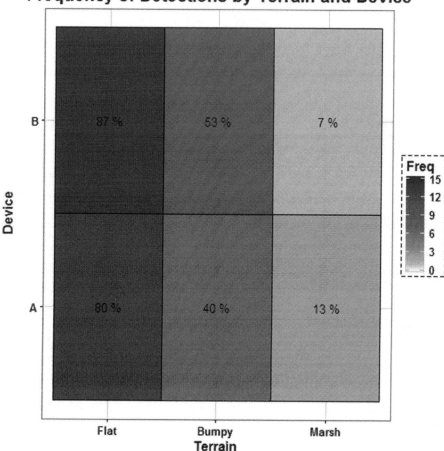

Frequency of Detections by Terrain and Device

This tile plot was only conducted on the Detections data and just gives an example of a tile plot larger than 2 × 2. Although with this data the bar chart is probably a clearer plot to use. These colors can always be amended dependent on the audience and media to be used

Tile plot colors can be changed depending on the data, for example, with the detections data clearly more detections is a good thing, hence green for higher values and red for lower values. However with the task completion data the completion and removals are both part of the variables, so a higher value doesn't necessarily mean a good thing, hence a more neutral color.

The first plot clearly shows that more participants Completed the course with Suit 1 and not Suit 2, and the second plot shows the clear separation between the three Terrains, but not much separation between the Devices.

Scatter Plot

Using the data from Chapter 7, Example 7.1, we can plot a scatter plot with a line of best fit showing the relationship between Yield and Concentration in Example 9.4.

The **geom_point()** is what allows us to create the scatter plot, then the **geom_smooth()** lets us add the line of best fit to the data. By choosing **"lm"** we are forcing the plot to fit a linear straight line to the data, which was chosen due to the fitted model output. The default in **geom_smooth()** adds standard errors to the line, which is shown by the gray area.

EXAMPLE 9.4

Creating a scatter plot in R using data from Example 7.1, data18, and adding a line of best fit.

```
ggplot(data18, aes(x = Conc, y = Yield)) + theme_bw() +
        geom_point() + geom_smooth(method = "lm") +
        xlab("Concentration") + ylab("Yield\n") +
        ggtitle("Yield by Concentration\n") +
        scale_y_continuous(limits = c(450,815),
                breaks= seq(450,815, by=50)) +
        theme(plot.title = element_text(size = 18, face = "bold"),
        axis.text = element_text(size = 10, face = "bold"),
        axis.title = element_text(size = 12, face = "bold"),
        panel.grid.major = element_line(colour = "grey90"))
```

Yield by Concentration

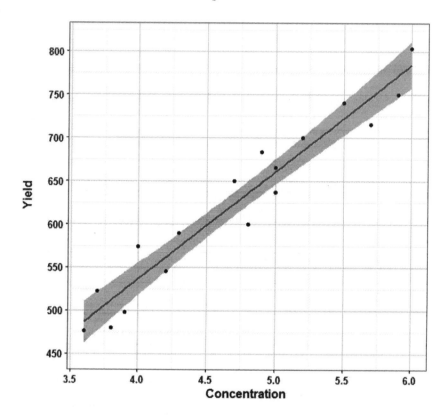

There are different methods for fitting the line of best fit, and you also can specify that it doesn't include the standard error region.

This clearly shows the positive relationship between Yield and Concentration. An additional piece of information that could be added to the plot is a reference line if there was a specific Yield required or a standard Concentration used.

Parallel Lines Plot

The parallel lines plot while not strictly a **ggplot()**, but instead requires the **PairedData** package, actually uses all the same aesthetics as a **ggplot()** would.

Using the data from Chapter 6, Example 6.8, we can plot a parallel lines plot for paired data to show the difference between the before training time and after training time for each participant in Example 9.5.

The **paired.plotMcNeil()** command required the data set, the two variables containing the paired data, and what the subjects will be; the variable doesn't have to be named Subject.

EXAMPLE 9.5

Creating a parallel lines plot in R using data from Example 6.8, data16.

```
library(PairedData)
windows(7,4)
paired.plotMcNeil(data16, "BeforeTraining", "AfterTraining",
            subjects = "Subject") + theme_bw() +
        scale_colour_manual(values = c("red", "blue")) +
        scale_x_continuous(limits = c(10,20), breaks = seq(10,20,
                by = 1)) +
        ggtitle("Time to Complete a Set Task Before and After
                Training\n") +
        xlab("Time to Complete a Set Task (mins)") +
        ylab("Subject\n") +
        theme(plot.title = element_text(size = 16, face = "bold"),
        legend.position = c(0.11,0.85),
        legend.background = element_rect(linetype = "dashed",
                colour = "black"),
        legend.title = element_text(size = 10, face = "bold"),
        legend.text = element_text(size = 10),
        strip.text = element_text(size = 10, face = "bold"),
        axis.text = element_text(size = 12, face = "bold"),
        axis.title = element_text(size = 12, face = "bold"),
        panel.grid.major = element_line(colour = "grey90"))
```

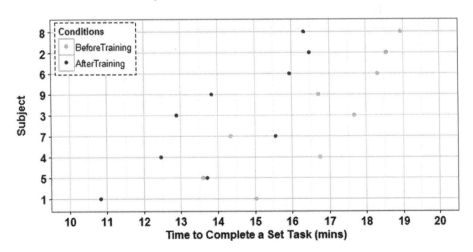

Time to Complete a Set Task Before and After Training

The plot automatically orders the subjects by a combination of the highest numbers mixed with the largest gap that at the moment unfortunately cannot be changed.

For any continuous paired data, these parallel lines plots are one of the best plots to use as they really emphasize the gaps within each participant.

Line Graph

Using data from a couple of the plots from Figure 9-5 earlier in the chapter, we can plot a line graph for the Temperature over Time per Location, and then pretend there was no difference between the Locations to plot the average Temperatures with confidence intervals in Example 9.6.

The **geom_point()** lets us define the points as before, then the **geom_line()** lets us add the lines connecting the points over Time—note that this will only work if the x-axis is ordered, such as with Time.

In the second plot the **geom_errorbar()** lets us add the confidence intervals that we could have ascertained using the **CI()** command in the **Rmisc** package mentioned in Chapter 5. The width defines the width of the confidence interval "ends."

EXAMPLE 9.6

Creating a line graph and a line graph with confidence intervals in R using data from Figure 9-5.

```
Temperature = c(43,55,47,64,60,53,57,50,60,62,50,58,56,55,58,57,60,
            58,58,56,64,64,53,53,60,67,66,62,48,55)
Day = rep(c(1:6), each = 5)
Location = rep(c("South East","South West","East of England",
        "The North", "Midlands"), 6)
data36 = data.frame(Temperature, Day, Location)
ggplot(data36, aes(x = Day, y = Temperature, colour = Location)) +
        geom_point(size = 2) + geom_line(size = 1) + theme_bw() +
        ggtitle("Temperature (°F) by Day and Location\n") +
        xlab("Day") + ylab("Temperature °F\n") +
        guides(colour = guide_legend(ncol = 2)) +
        scale_y_continuous(limits = c(42,68), breaks = seq(42,68,
            by = 2)) +
        scale_x_continuous(limits = c(1,6), breaks = seq(1,6,
            by = 1)) +
        scale_colour_manual("Location", values = c("firebrick3",
                "forestgreen","dodgerblue3","orange","purple")) +
        theme(plot.title = element_text(size = 18, face = "bold"),
        legend.position = c(0.77, 0.11),
        legend.background = element_rect(linetype = "dashed",
                colour = "black"),
        legend.title = element_text(size = 12, face = "bold"),
        axis.text = element_text(size = 10, face = "bold"),
```

```
        panel.grid.major = element_line(colour = "grey90"),
        axis.title = element_text(size = 12, face = "bold"))
```

Temperature (°F) by Day and Location

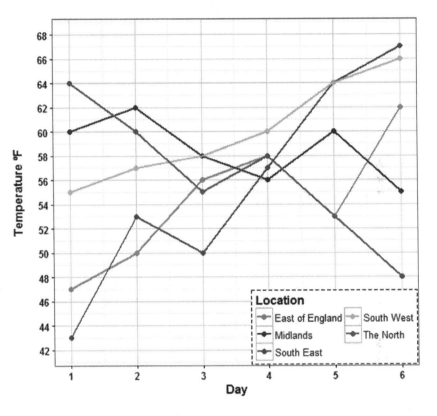

```
xbar = c(53.8,56.4,55.4,57.8,58.8,59.6)
Day = c(1:6)
LCI = c(42.93,50.28,51.32,55.96,51.92,49.64)
UCI = c(64.67,62.52,59.48,59.64,65.68,69.56)
data37 = data.frame(xbar, Day, LCI, UCI)
ggplot(data37, aes(x = Day, y = xbar)) + geom_line(size = 1) +
        geom_errorbar(aes(ymin = LCI, ymax = UCI), width = 0.25,
               size = 1) +
        geom_point(size = 3, pch = 21, fill = "white") +
        theme_bw() + ggtitle("Average Temperature by Day
        with 95% Confidence Intervals\n") +
        xlab("Day") + ylab("Average Temperature (°F)\n") +
        scale_y_continuous(limits = c(42,70), breaks = seq(42,70,
               by = 2)) +
        scale_x_continuous(limits = c(0.75,6.25), breaks = seq(1,6,
               by = 1)) +
```

```
theme(plot.title = element_text(size = 18, face = "bold"),
axis.text = element_text(size = 10, face = "bold"),
panel.grid.major = element_line(colour = "grey90"),
axis.title = element_text(size = 12, face = "bold"))
```

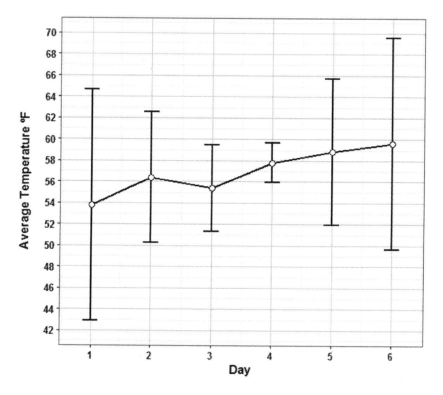

For the first plot you could add in different line types as well as or instead of different colors. In addition the legend can be moved around to suit the plot better.

The second plot made the choice to fill the points in white just to be able to highlight where that average temperature lies. The *y*-axis label and the title also have been amended to clearly define that these are average temperatures.

There should never be a plot of averages or medians without some sort of confidence interval or error bars. However thought should be given as to whether summarizing the data is the best option and testing for differences between the groups should always be carried out first.

Box Plot

Using the data from Chapter 7, Example 7.3, we can plot a box plot to show the difference between the concentrations and machines in regards to time in Example 9.7.

The *geom_boxplot()* lets us plot the box plot and define what our statistical outliers should look like. However *stat_boxplot()* lets us add the "ends" to the "whiskers" on the boxes.

The *stat_summary()* command allows us to add extra information to the plot, in this case by adding a mean point to see how it differs to the median values.

EXAMPLE 9.7

Creating a box plot in R using data from Example 7.3, data20.

```
ggplot(data20, aes(x = Concentration, y = Time.Taken)) + theme_bw() +
        facet_wrap(~ Machine) + stat_boxplot(geom = "errorbar") +
        geom_boxplot(outlier.colour = "red", outlier.shape = 7,
                outlier.size = 1.5) +
        ggtitle("Time Taken by Concentration \n and Machine\n") +
        xlab("Concentration") + ylab("Time Taken (Mins)\n") +
        stat_summary(fun.y = mean, geom = "point", shape = 8,
                size = 2, col = "blue") +
        scale_y_continuous(limits = c(25,50), breaks = seq(25,50,
                by = 5)) +
        theme(plot.title = element_text(size = 16, face = "bold"),
        strip.text = element_text(size = 12, face="bold",
                lineheight = 3),
        axis.text = element_text(size = 12, face = "bold"),
        axis.text.x = element_text(size=9, face="bold", angle=45,
                h = 1),
        axis.text.y = element_text(size = 9, face = "bold"),
        axis.title = element_text(size = 12, face = "bold"),
        panel.grid.major = element_line(colour = "grey90"))
```

Time Taken by Concentration and Machine

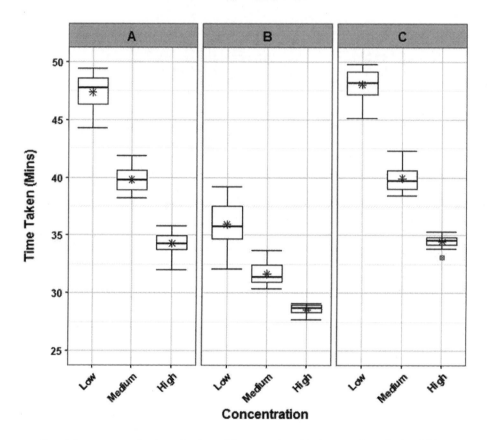

Box plots are an excellent way to show differences found between categorical variables during statistical analysis. They also are very handy for showing the variation within each group and whether the mean and median are at similar values.

Just to reiterate that just because a box plot doesn't overlap with another box plot, this does not mean there will be a significant difference between the groups; although likely, it very much depends on the sample size.

Likert Plot

There are a couple of ways to plot a Likert plot for ordinal data, however I prefer to use the version within the **HH** package, so again not a **ggplot()**.

Using the data from Chapter 7, Example 7.9, we can plot a Likert plot to show the spread of responses to a specific question over different systems in Example 9.8.

The initial section of the Likert plot is set up similar to a model, however the left side of the model is what should be plotted on the y-axis and isn't actually the response variable. If you have the frequencies of the responses, then you will need to define that here **value = "Responses."**

You can define that the results be shown in percentages as well as showing the sample size in each group by letting **as.percent = TRUE**.

You can also play with the **ReferenceZero** value to change where the middle line should be. This is more relevant when there is a neutral category as the reference line can either be in the middle of this, or the neutral category can fall into the "negative" or "positive" side of the plot.

EXAMPLE 9.8

Creating a Likert plot in R using data from Example 7.9, data28.

```
library(HH)

windows(7,5)
plot.likert(System ~ Likert.Response|Group, value = "Response",
        data = data28, layout = c(1,2), as.percent = TRUE,
        ReferenceZero = 2.5,
        main = "How Poorly/Well did the System Perform
                with your Current Equipment?",
        xlab = expression(bold("Percent")),
        ylab = expression(bold("System")),
        col = c("firebrick3","indianred1","springgreen",
                "forestgreen"),
        scales = list(x = list(limits = c(-102,102),
                at = seq(-100,100,10), labels = abs(seq(-100,
                100, 10)), cex = 0.65, tck = 0.5)))
```

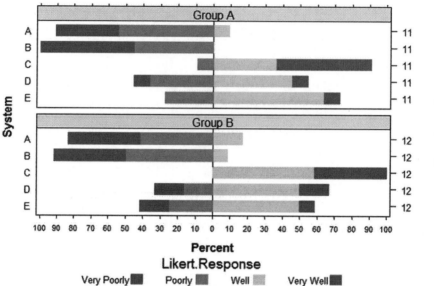

How Poorly/Well did the System Perform with your Current Equipment?

Likert plots are really good at displaying ordinal data, as there will be appropriate coloring for the ordered data. In addition to be able to define the "middle" allows for more control over neutral categories and whether they are viewed as positive, neutral, or negative.

There are some preset color patterns that can be used. They are from the **RColourBrewer** package, which should automatically have been loaded with other packages such as **ggplot2**. In Example 9.8 the list of colors could be replaced with:

col = brewer.pal(4, "RdYlGn").

This would tell it to look at the colors in the **"RdYlGn"** set and chose four of them, the number of Likert responses, to use in the Likert plot. There are other color sets though they generally have a limit to how many Likert responses there can be.

Trellis Plot

Using the data from Chapter 7, Example 7.10, we can plot a trellis plot to show the trend of heart rate over time for two treatments per participant in Example 9.9.

The **geom_point()** lets us plot the data, which will be both colored and shaped by the treatment variable.

EXAMPLE 9.9

Creating a trellis plot in R using data from Example 7.10, data29.

```
ggplot(data29, aes(x = Time, y = Heart.Rate, colour = Treatment,
        shape = Treatment)) + theme_bw() + geom_point() +
        facet_wrap( ~ Subject) + xlab("Time (Secs)") +
        ylab("Heart Rate\n") +
        ggtitle("Heart Rate by Time and Treatment
                per Subject\n") +
        scale_y_continuous(limits = c(0,201), breaks = seq(0,200,
                by = 50)) +
        scale_colour_manual("Treatment", values = c("firebrick3",
                "dodgerblue3")) +
        scale_shape_manual("Treatment", values = c(3,4)) +
        theme(plot.title = element_text(size = 16, face = "bold"),
        legend.position = c(0.75,0.18),
        legend.background = element_rect(linetype = "dashed",
                colour = "black"),
        legend.title = element_text(size = 10, face = "bold"),
        legend.text = element_text(size = 10),
        strip.text = element_text(size = 10, face = "bold"),
        axis.text.x = element_text(size = 9, angle = 45,
                hjust = 0.5, vjust = 0.5),
        axis.text.y = element_text(size = 9, face = "bold"),
        axis.title = element_text(size = 12, face = "bold"),
        panel.grid.major = element_line(colour = "grey90")))
```

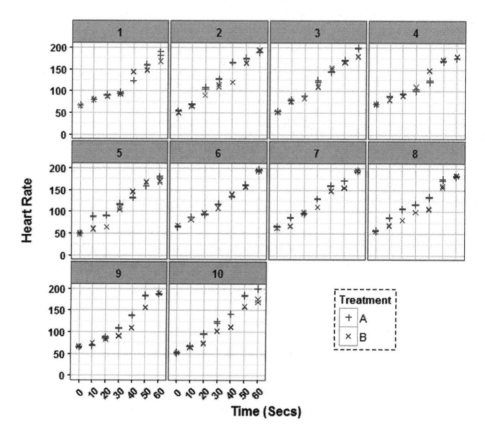

By grouping the data within participants we can see the relative differences between the treatments, as the data is not independent.

Trellis plots are really easy to create but can show a lot of information. The trellis plot can be created for any type of **ggplot()**, for example, box plot, scatter plot, and so forth.

Logistic Curve

Using the data from Chapter 7, Example 7.7, we can plot a logistic curve to show the probability of detection by concentration in Example 9.10.

To plot the curve some analysis should have already been completed, which was done in Example 7.7. This requires predicting values based on the fitted model and assigning them to a new dataset, here called ***tmpModFit***.

The ***geom_point()*** lets us plot the raw binary data, then the ***geom_line()*** allows us to add in the predicted probability curve as well as adding in the 95% confidence intervals around this later in the code.

The ***geom_segment()*** lets us add in a reference line, here at a probability of detection of 0.95; one from the y-axis to the curve, and one from the curve to the x-axis. The corresponding concentration can be found out using the ***dose.p()*** command in the ***MASS*** package, as was done in Example 7.7.

The ***annotate()*** command allows us to add in text to explain the reference line.

EXAMPLE 9.10

Creating a logistic curve in R using data from Example 7.7, data25. The whole of Example 7.7 will need to be run first to calculate the curve and confidence intervals to be used in the plot.

```
windows(6,6)
ggplot(data25, aes(Concentration, Detection)) + geom_point() +
        geom_line(aes(Concentration, Proportion), tmpModFit) +
        xlab("Concentration") + ylab("Probability of Detection\n") +
        ggtitle("Probability of Detection
                by Concentration (95% CI)\n") +
        guides(colour = FALSE) + theme_bw() +
        geom_segment(aes(x = 0, y = 0.95, xend = 368, yend = 0.95),
                colour = "blue", linetype = "dotted") +
        geom_segment(aes(x = 368, y = 0.95, xend = 368, yend = 0),
                colour = "blue", linetype = "dotted") +
        annotate("text", x = 447, y = 0.1, colour = "blue",
                label = "Average \n Concentration for
                95% Detection") +
        geom_line(aes(lowerv50, prop), tempv50, colour = "red",
                linetype = "dashed") +
        geom_line(aes(upperv50, prop), tempv50, colour = "red",
                linetype = "dashed") +
        scale_x_continuous(limits = c(0,500), breaks = seq(0,500,
                by = 50)) +
        scale_y_continuous(limits = c(0,1), breaks = seq(0,1,
                by = 0.1)) +
        theme(plot.title = element_text(size = 18, face = "bold"),
        axis.text = element_text(size = 10, face = "bold"),
        axis.title = element_text(size = 12, face = "bold"),
        panel.grid.major = element_line(colour = "grey90"))
```

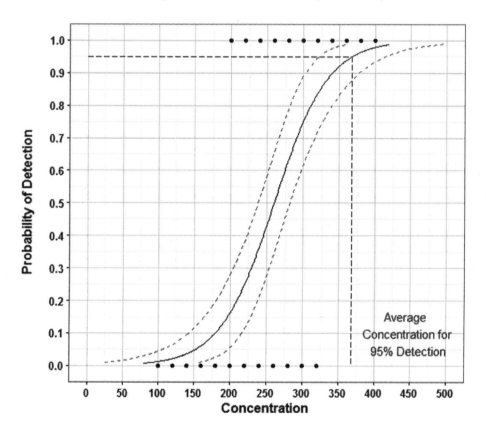

The logistic curve needs to be fitted with predictions from the final model, and it's always advisable to include confidence intervals to show the variability. There may be very little data at one end so the confidence interval would then be very wide. A reference line can be quite helpful if there is a threshold or a requirement for a certain probability of detection.

These curves also can be known as logit or probit curves. This is just a link option when fitting the model with the default being logit. They both produce very similar results, but generally the probit link is chosen if the data approaches zero or one quite rapidly.

Summary

This chapter was split into three main sections, general tips as to what to avoid when creating plots, how to amend the aesthetics in R plots, and finally common plot types in R and how to create them.

Within the general tips some common plotting mistakes were described such as using 3D unnecessarily, incorrectly plotting averages, using plots when a simple table would be preferable, and more. Chapter 3 covered the information about which plot type should be used for which data types so it was not repeated here; though using the wrong plot for the data is one of the most common mistakes.

The second section gave an example of a plot created in R without adding in any aesthetic details, then creating the plot again with these details. It then went on to discuss what each section of the R code controls and how to amend it for future plots.

The final section looked at some of the more common plots, mainly using data and examples from previous chapters, including how to create them in R including all the aesthetic details. It showed which sections were the main commands to create the plots along with other details that could be included within them.

Chapter 10 delves into translating and communicating results to customers. Advice about this has been dotted throughout the book, especially at the end of examples where the results have been summarized, however Chapter 10 ties all this information together.

Translation and Communication

How Do I Get the Message Across?

The most important thing, apart from conducting the right analysis on the right data, is to be able to translate the results to the customer. Although this has been done through all the examples in the previous chapters, this chapter pulls everything together.

The chapter is divided into two sections, the first includes general advice, and the second a few examples of suggested ways to present the information.

General Advice

The following includes some issues that can arise when attempting to get the message across as well as some general guidance for translation.

Review

Analysts can get wrapped up in both the subject they deal with and the statistical methodology they use. Therefore, they may not write the results as clearly as possible, which will be picked up in the review process.

© Victoria Cox 2017
V. Cox, *Translating Statistics to Make Decisions*, DOI 10.1007/978-1-4842-2256-0_10

In addition, depending on the size of the project undertaken, my advice is always to first run your planned analysis by another statistician. There is nothing worse than finding out you haven't chosen the right method, or that there may be a new development you have not heard about.

There should be at least two reviewers involved at the final stage, both on the subject matter side and the statistical side. The subject matter expert can check that the assumptions and conclusions are viable in a realistic sense while the statistician can check that the data has been treated correctly and the conclusions match the analysis output.

Provide the Right Level of Information

Providing a sufficient level of information to the customer needs to be a balance of getting the key messages across without excluding assumptions or adding too much detail.

The best way to do this is to have a separate summary of the analysis that can be read independently of other reports, then to add a technical report either as an annex to a project report or as a standalone report. The summary of analysis should fit on a side or two of US letter (A4 paper), which depends on the size of the plots, as they take up the most space.

The summary of analysis should include a brief introduction of the data along with why the analysis was being carried out, an overview of the more important assumptions plus the sample size used, the key messages found from the analysis along with relevant visual aids, and any recommendations. An advantage to the short summary of analysis is that information shouldn't get lost along the way, sometimes if you provide a technical report, someone else will pick the key messages out to include in a project report that can lead to them being quoted incorrectly or assumptions being left out.

The technical report can then include all the analysis carried out; although I would still never include actual software output, I would always translate it. There also should be a detailed methodology section so it is clear what type of analysis was run on which sections of the data. In addition if R was used, the version and the packages installed should be listed.

Include Assumptions

The analysis conducted will not apply for all cases and scenarios, there clearly will be limitations to the conclusions and as such, they need to be listed. There will be varying degrees of importance for each of the assumptions, and these should be clear within your own study. For example, assumptions about the type of scenario the results can be applied to are quite important, whereas assumptions about the distribution of the data are much less so in the eyes of the customer.

Prioritize the assumptions and list only the important ones in the summary of analysis. Linked closely to this are the limitations; for example, if confounding has occurred it needs to be specified as you can't clearly state that the effect is down to one variable over another.

Adapt Language

The language should be adapted to suit the customer. You may need to use different language when reporting to the military as opposed to a CEO of a company or compared to an academic partner. You should be able to determine this through conversations throughout the project with the customer.

It may be just at the formal/informal writing level, the use of abbreviations, or it could even be at the details level. For example do they prefer stating evidence of a significant difference, at a specific percentage confidence, or do they actually require p-values to be shown?

Give the Correct Answer

There are two pitfalls that people can fall into here, either letting themselves be bullied into giving the "right" answer or not wanting to show a "negative" answer.

Occasionally a decision may have already been made about an option or piece of equipment, or it may just be that they have a favorite in mind. However, it's always a good idea to keep the analyst independent of this as it helps with the robustness of the analysis. However, in these cases you can have pressure put on you to show the result that they want to see. Although it can be difficult you shouldn't let yourself be swayed and should quote the conclusions as they are found. It helps if you don't know the preferences as then you can't be biased in either direction. I also would push to be a reviewer of any final project report, this way you can check that they haven't misquoted the analysis or purely excluded it.

Linked to this is giving a negative answer, especially if a lot of money has been spent on a project. A negative answer is still a valid answer. This especially holds when looking at new research ideas; it may be that the item or idea isn't feasible for what was intended. This answer is just as important as if it had worked; in fact I think more "negative" answer papers should be published. Showing a negative answer can prevent future funds being wasted by pursuing the idea, it can highlight whether the idea was completely wrong, or if it just needs tweaking. It's this final point that needs to be conveyed to the customer, it's not that no work was conducted; it's that the research showed that the product/idea/methodology wasn't appropriate and so future funds should be placed elsewhere.

Translate the Answer

Quite simply, make sure the answer has been translated to English. If the answer is left in "statistics speak" then people could disengage or may not understand the results.

Make sure the emphasis is in the subject matter language and that units are reintroduced to give more of an idea about the size of the effect or the level of uncertainty in the results.

Examples of Translation

Using examples from previous chapters, I provide summaries of analyses from previous examples to highlight the length required to actually get the key messages across, how to translate the statistical output into English, and how to include assumptions.

I look at four examples of varying complexity from Chapters 5, 6, and 7. No analysis will be shown in the examples as this has already been undertaken in the chapters, the only information will be the theoretical discussion with the customer and the final translation of the key messages.

Translation Example 1

Let's say that the data provided in Example 5.1, data7, are the peak operating temperatures for a piece of machinery. The results quoted are pulled from Example 5.1 and Example 5.6, which calculated confidence and tolerance intervals.

The initial question from the customer was "how variable is the peak operating temperature?" This clearly needed better definition to be answered properly, so through discussions both confidence intervals and tolerance intervals were explained in practical terms.

- A confidence interval shows the uncertainty around the average peak temperature.

- A tolerance interval shows the uncertainty around a proportion of peak temperatures, such as 90%.

The customer was interested in both statistics for differing reasons, so the next step was to explain risk so we knew which confidence and coverage to use in the calculations.

- How much risk are you willing to accept that the interval doesn't contain the true average peak temperature? Then this value can be turned into a confidence level and also can be used for the tolerance interval.

- What proportion of total peak temperatures should fall within the interval?

The time and cost limitations meant that a sample size of 18 could be recorded. The customer decided on a confidence level of 90% and a coverage level of 75%.

The results were as follows:

- We are 90% confident that the average peak operating temperature is between 27.1°C and 27.7°C.

- We are 90% confident that at least 75% of peak operating temperatures are within 26.2°C and 28.5°C.

- The data was assumed to follow a normal distribution after visual examination. A confidence level of 90% means there is a 10% risk that the intervals do not contain the required value of peak temperature—either the average or the proportion of 75%. A coverage level of 75% means that 25% or less of the peak temperatures will be outside the bounds.

This made the uncertainty around the peak operating temperatures very clear to the customer and due to the results showing little variation, no further action or testing needed to be undertaken.

As a side note, no plot was necessary for this output as it would not have made the key messages any clearer.

Translation Example 2

We use Example 6.2–Table, which concerns whether there was a difference between two treatments in terms of curing an ailment.

As this treatment is a drug the customer wants to work at a confidence level of 99%. One of the treatments is the current treatment, but to keep impartiality the identity of the current treatment has not been disclosed. Therefore the hypothesis test to be carried out will look for a difference in either direction rather than just seeing if the new treatment is better than the current treatment.

Due to cost the risk of a false negative, the risk that no significant difference is found when there should have been one, was decided to be a lower priority and was chosen at 20%. This meant that the power was 80% that led to a sample size of 30 per group, 60 total being acceptable for the study, which also satisfied the ethical review.

The results were as follows:

- The data was recorded in binary form, either the subjects ailment was cured or it wasn't. There were also different subjects in each group to eliminate cross-over effects.

- There was no evidence of a significant difference between the treatments at the 99% confidence level (p-value of 0.026). Treatment A cured more than Treatment B in the sample, 83.3% compared to 53.3%, respectively, see Figure 10-1.

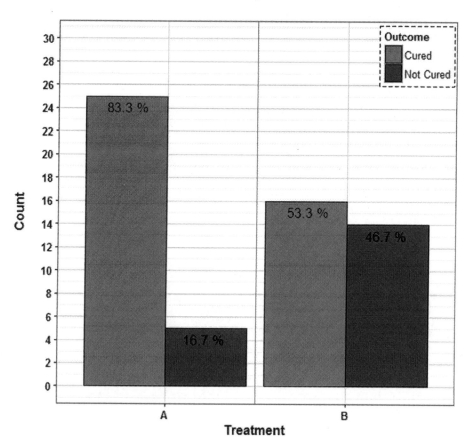

Figure 10-1. Plot for Translation Example 2

- The size of the effect or difference between the treatments was 30%, however the difference could be as large as 62.6% or as small as 0.03%, which means Treatment B could cure more than Treatment A.

It transpired that Treatment A was the current treatment, which did actually produce significantly better results at the 99% confidence level, so no action needed to be taken.

An additional piece of information that should be included in a technical report is the methodology used and its justification.

Translation Example 3

We use Example 7.3, data20, which looked at the time taken to find a substance of interest in terms of machine, operator, and concentration.

The sample size for each factor combination is 7, however if there are no significant interactions, this soon increases. We believed that there was no difference between Operators, however it was included for completeness but without a difference the sample size was 14.

The results were as follows:

- The data was approximately normally distributed so a linear model was run with the resulting model being a good fit explaining 96.1% of the variation.

- The final model showed that Machine, Concentration, and the interaction between Machine and Concentration was significant at the 99% confidence level.

- Machine B gave significantly quicker time taken than both Machine A and Machine C at the 99% confidence level.

- High Concentration gave significantly quicker time taken than both Medium and Low Concentration at the 99% confidence level. Medium Concentration gave significantly quicker time taken than Low Concentration at the 99% confidence level.

- In general terms, the interaction between Machine and Concentration was significant as the downward gradient for Machine B by Concentration was less extreme than those for Machine A and Machine C, see the plot in Figure 10-2 and the technical report for more details.

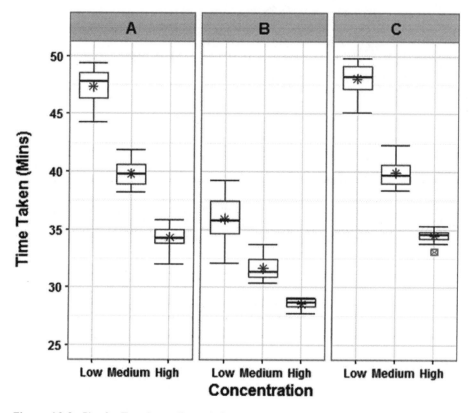

Figure 10-2. Plot for Translation Example 3

- The model estimates for varying Machines and Concentrations are shown in Figure 10-3.

Model Estimates		Concentration		
		Low	Medium	High
Machine	A	47.35	39.81	34.31
	B	35.92	31.65	28.55
	C	48.04	39.92	34.47

Figure 10-3. Model Estimates for Translation Example 3

In this study the differences between the Concentrations were expected, however the differences between the Machines were not expected so further investigation will be given as to why Machine B gave such different results than Machine A and Machine C.

Translation Example 4

This example translates the output from Example 7.7, data25, which investigated the number of detections at different concentrations.

The initial design of experiments used the Staircase method to narrow the concentration area of interest to between 100 and 400; measurements were then taken at every 20 units with 6 repeats at each as agreed with the customer.

The results were as follows:

- As there were Concentrations that resulted in all non-detects or detects, a bias-reduction binomial-response generalized linear model (BRGLM) was used to account for this.

- The model showed that Concentration was significant at the 99% confidence level—Figure 10-4 shows the curve of Probability of Detection by Concentration.

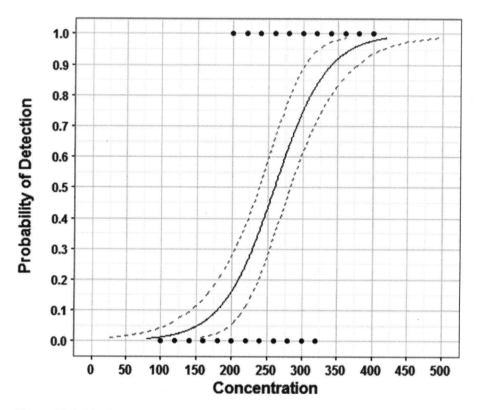

Figure 10-4. Plot for Translation Example 4

- For a one unit increase in Concentration the odds ratio of Detection is 1.028, with a 95% confidence interval of 1.019 to 1.042

- The required Concentration level with 95% confidence intervals for varying Probabilities of Detection is shown in Figure 10-5.

Probability of Detection	Concentration	Confidence Interval (95%)
0.50	260.19	+/- 21.94
0.75	300.42	+/- 26.78
0.90	340.64	+/- 37.41
0.95	368.00	+/- 46.02
0.99	428.44	+/- 66.62

Figure 10-5. Required Concentrations for Translation Example 4

The last figure in this example clearly shows the required concentrations to give a certain probability of detection, and while this is the most important piece of information for the customer, the model needed to be a good fit to produce useful values.

Summary

This chapter was split into two sections, general tips for what to avoid and what to ensure with translation and communication of results, then four examples using data from previous chapters.

Within the general tips some common pitfalls are discussed such as including unnecessary information in the summary, not undertaking a sufficient level of review, and trying to coerce the results to show the "right" answer. Whether that's what the customer has been pushing for or to avoid showing a "negative" answer. It also highlights some guidance about including assumptions, adapting the language for the audience, and making sure the results are translated from the statistical jargon into English.

The second section looked at four translation examples to highlight that it doesn't take much space to emphasize the key findings to the customer, and that adding figures and tables also can visually help to improve understanding of the results. Also a reminder to make sure any main assumptions are made clear up front, such as those including confounding variables, outliers, and state the sample size. Any details of analysis, explaining the methodology and other information can be and should be included in a separate technical report. In addition I also would include the "key messages" part at the beginning of the technical report.

Hopefully, this book has shed some light on the confusing world of statistics, emphasizing that while the analysis needs to be conducted correctly, the final output should be as clear and concise as possible. It has delved through the general process for a statistical study beginning with the design of experiments and data collection; through the exploratory data analysis, descriptive statistics, measuring uncertainty, and simple hypothesis testing; on to the more complex statistical modeling and multivariate analysis; and then finally how to produce detailed plots and translate the output to a nonstatistically minded customer. It has looked at statistical analysis from the viewpoint of someone carrying out the analysis, however it also has highlighted the key issues and pitfalls that those not conducting the analysis should be looking out for in addition to explaining unknown or previously baffling statistical terms.

I

Index

A

Accept the null, 126

Advice, 307, 308

Aesthetics, 269, 279–285

Alternative hypothesis, 126

Analysis of variance (ANOVA), 176–183

Arithmetic mean. *See* Location, mean

Assumptions, 11

Autocorrelation, 163

B

Bartlett's test, 146

Bias-reduction binomial-response GLM (BRGLM), 207–210

Binomial GLM, 202–206

Bivariate data, 99–101

C

Categorical data. *See* Qualitative data, nominal data

Chi square test, 134

Clopper–Pearson interval. *See* Confidence interval, exact method

Coefficients, 226, 228, 230, 238

Color, 274, 277–279, 283–285

Communication, 307–318

Computer experiments
deterministic, 25
Latin hypercube sample (LHS), 25–28
low discrepancy sequence, 25, 26
stochastic, 25
uniform design, 26–28

Confidence interval
asymptotic normal interval, 112–113
exact method, 112–113
Wilson interval, 112

Confounding, 14–16

Contingency tables, 99

Correlation, 99–100

Covariance, 99–100

D

Data collection, 33–34

Descriptive statistics, 76

Diagnostic plots, 163–165, 174, 180, 183, 191, 193, 226

3D, 272–273

Dispersion parameter, 210

Distribution
binomial, 70
exponential, 68, 69
gamma, 68, 69
Gaussian, 65
Poisson, 71

© Victoria Cox 2017
V. Cox, *Translating Statistics to Make Decisions*, DOI 10.1007/978-1-4842-2256-0

Get the eBook for only $4.99!

Why limit yourself?

Now you can take the weightless companion with you wherever you go and access your content on your PC, phone, tablet, or reader.

Since you've purchased this print book, we are happy to offer you the eBook for just $4.99.

Convenient and fully searchable, the PDF version enables you to easily find and copy code—or perform examples by quickly toggling between instructions and applications.

To learn more, go to http://www.apress.com/us/shop/companion or contact support@apress.com.